Dancing on the Tails of the Bell Curve

Dancing on the Tails of the Bell Curve
Readings on the Joy and Power of Statistics

Edited by

Richard Altschuler

Gordian Knot Books
An Imprint of Richard Altschuler & Associates, Inc.

Los Angeles

Distributed by University Press of New England
Hanover and London

BOWLING GREEN STATE
UNIVERSITY LIBRARIES

Dancing on the Tails of the Bell Curve: Readings on the Joy and Power of Statistics. Copyright © 2013 by Richard Altschuler. For information contact the publisher, Gordian Knot Books, at 10390 Wilshire Boulevard, Los Angeles, CA 90024, (424) 279-9118, or send an email to Richard.Altschuler@gmail.com.

Library of Congress Control Number: 2013914719
CIP data for this book are available from the Library of Congress

ISBN-13: 978-1-884092-91-6

Gordian Knot Books is an imprint of
Richard Altschuler & Associates, Inc.

The essays and articles in this anthology are published with the permission of the authors or journals in which they originally appeared or are in the public domain. All rights reserved. Except as allowed for fair use, no part of this publication may be reproduced, stored in a retrieval system, or transmitted in any form or by any means, electronic, mechanical, photocopying, recording, or otherwise, without the prior written permission of Richard Altschuler & Associates, Inc.

Cover Design: Josh Garfield

Printed in the United States of America

Distributed by University Press of New England
1 Court Street
Lebanon, New Hampshire 03766

To Jane
the joy and power in my life
who has always been my dancing partner
and forever will be

Table of Contents

Prologue: U. S. Congressional Resolution to Designate
2013 as the "International Year of Statistics"   1

Editor's Introduction   4

The "Godliness" of Statistics * Florence Nightingale   14

The Magnificence of the Mean *Adolphe Quetelet   17

Probability: The Key to Reason and Religion
* Thomas Bayes and Simon Laplace   22

The "Charms" of Statistics * Sir Francis Galton   26

How Statistics Helped Make Biology a Science
* Karl Pearson and *Biometrika*   30

Darwin: The Reluctant Mathematician by Julie Rehmeyer   35

Statistical View of the United States by D. B. Debow   39

The Importance of Statistics to the United States Census
and the World by President James A. Garfield   41

The First Hundred Years of the Bureau of Labor Statistics
(1884-1984) by Joseph P. Goldberg and William T. Moye   47

Statistics and Government by Wesley C. Mitchell   52

How Statistics Helped Desegregate America's Public School
System * Kenneth Clark and the U. S. Supreme Court   55

The 1964 Surgeon General's Report on Smoking   61

2013: The International Year of . . . Statistics
by Marie Davidian   65

BLS 125th Anniversary   71

The Necessity of Statistics in Highway Construction
Management: The Case of Arkansas by Harold Rothbart   75

Could a Statistic on Redistricting Have a Game-Changing
Effect on American Politics? by Thomas R. Belin   81

The Importance of "Ag" Estimates   87

Data Inquiry and Analysis for Educational Reform
by Howard H. Wade   89

Interview with Francesca Dominici   97

The Joy of Research Discovery with Statistics
by Allan Geliebter   102

Smoking in Movies by Tim McAfee and Michael Tynan   106

Statistics, *Moneyball* and "Sabermetrics"   111

Data-Driven Equity in Urban Schools by Wendy Schwartz   114

Statistics and the Fashion Industry   119

Epilogue: The Joy of Perceiving that Basic Statistics Is a
"Mean World" by Richard Altschuler   128

A Potpourri of Quotations about Statistics and Statisticians   136

Works Cited   145

PROLOGUE

The year 2013 was recognized as "The International Year of Statistics" by over two thousand organizations worldwide that celebrated and honored the discipline of statistics. They included the U.S. Congress with the following Resolution, introduced by Senator Kay Hagan (D-NC), which details the many ways the science, art, and profession of statistics benefit individuals and nations.

113th CONGRESS
1st Session
S. RES. 150

TO DESIGNATE THE YEAR 2013 AS THE
"INTERNATIONAL YEAR OF STATISTICS"

IN THE SENATE OF THE UNITED STATES

May 21, 2013

Resolution

To designate the year 2013 as the 'International Year of Statistics'.

Whereas more than 2,000 organizations worldwide have recognized 2013 as the International Year of Statistics, a global celebration and recognition of the contributions of statistical science to the well-being of humankind;

Whereas the science of statistics is vital to the improvement of human life because of the power of statistics to improve, enlighten, and understand;

Whereas statistics is the science of collecting, analyzing, and understanding data that permeates and bolsters all sciences;

Whereas statisticians contribute to the vitality and excellence of myriad aspects of United States society, including the economy, health care, security, commerce, education, and research;

Whereas rapidly increasing numbers of students in grades K through 16 and educators are recognizing the many benefits of statistical literacy as a collection of skills to intelligently cope with the requirements of citizenship, employment, and family;

Whereas statisticians contribute to smart and efficient government through the production of statistical data that informs on all aspects of our society, including population, labor, education, economy, transportation, health, energy, and crime;

Whereas the goals of the International Year of Statistics are to increase public awareness of the power and impact of statistics on all aspects of society, nurture statistics as a profession, especially among young people, and promote creativity and development in the sciences of probability and statistics; and

Whereas throughout the year, organizations in countries across the world will reach out to adults and children through symposia, conferences, demonstrations, workshops, contests, school activities, exhibitions, and other public events to increase awareness of the history and importance of statistics: Now, therefore, be it

Resolved, That the Senate —

(1) designates the year 2013 as the 'International Year of Statistics';

(2) supports the goals and ideals of the International Year of Statistics;

(3) recognizes the necessity of educating the public on the merits of the sciences, including statistics, and promoting interest in the sciences among the youth of the United States; and

(4) encourages the people of the United States to participate in the International Year of Statistics through participation in appropriate programs, activities, and ceremonies that call attention to the importance of statistics to the present and future well-being of the people of the United States.

Source

Congressional Record. (2013). May 21, page S3662. http://thomas.loc.gov/cgi-bin/query/z?c113:S.RES.150

Editor's Introduction

How I Became Motivated to Learn and Use Statistics

When I was an undergraduate student, I had no interest in learning statistics. The same was true when I began graduate school. I had taken the required "stat" courses to that point, and struggled very hard to get a "B" in them. I cannot say I had much enjoyment throughout that struggle, for I had to put in many anxiety-filled hours of calculating values on a hand-held electronic calculator, learning what seemed to me like arcane concepts, and answering "word problems," which often seemed to have little or no relevance to the outside world, the "real world."

But my attitude toward statistics changed when I met a person who *inspired* me to want to learn and use statistics. In fact, he did even more than this: He exuded *joy* when he stood before his blackboard, with his sleeves rolled up to his elbows, working out the solutions to statistical problems. When I was in his office alone with him, he would go through these different exercises, trying to instill in me the joy and power that statistics held for him, and their importance in the world at large.

As a result of his enthusiasm, seeing the pleasure he got from statistics, I suddenly found I had the *desire* to want to learn and use statistics. It was a good thing, too, because I had just been offered an assistantship to teach statistics to new students: Stat 101! And this was when I was in the master's degree program, as a sociology major at Temple University in Philadelphia, PA., at a time when I had to struggle so hard to get through my stat courses with a "B."

That professor, who "turned me on" to stat, was named Robert H. West, and he eventually became an Associate Dean at Temple University. I am certain that if I did not have a role model in "Bob"

West, I never would have been motivated to really *want* to learn how to do statistics, understand the concepts, use them in real life, in whatever ways I could for the good of myself and others. As it turned out, throughout my life I have personally been involved intimately with statistics in many ways—as a professor, research study consultant, tutor, data analyst, citizen, and consumer, among many other ways.

Ever since those private meetings with Bob West, I have had the feeling in the back of my mind that it would be a great idea to assemble a book of essays that would convey the joy people experience from working with statistics and describe how statistics have influenced the nature and course of our society and world in important ways.

This book is the result of that impulse, and I hope it does for you what the late Bob West did for me, in his little office at Temple University, several decades ago. That is, I hope it *inspires* you to want to learn statistics, and *motivates* you to want to use them in your personal and professional life.

The Value of Motivation for Learning

As you undoubtedly know, before we can accomplish anything great or important in life we need proper *motivation* or *inspiration*. Whether we want to lose weight, advance our careers, be great parents—or learn statistics!—we can only go so far, get into what we are doing so deeply, and no more, without proper motivation or inspiration. What's more, without proper motivation, our work can seem like a struggle, a great effort, an uphill climb all the way, usually with minimal success and a lot of complaining along the way. With proper motivation or inspiration, however, the hardest tasks can seem like "child's play," and even seem like fun, a great deal of the time—because that is the power of *wanting* to do something, rather than *having* to do something, of *understanding* why it is good or important to do something, rather than *feeling forced* to do something in life without understanding why.

So I hope that you find it within yourself to become motivated and inspired to learn statistics, to use statistics, for the good of society and yourself. This book will hopefully spur you in that direction, and perhaps you will also find a mentor, as I did, to "turn you on" to the joy and power of statistics. There is every reason to want to do that today and nothing to stop you from being expert, or at least competent, in basic statistics, and achieving a level of statistical literacy you never thought possible.

People and Events You Will Learn about in this Book

From reading the Prologue—about the honor the United States Congress bestowed upon the science of statistics, with its "Resolution" in conjunction with the "International Year of Statistics: 2013"—I hope you have already seen the importance of statistics in our society. And I hope that your feeling is magnified as you continue to read the essays and articles in this book about people and events that convey the joy and power of statistics.

As you'll see, for example, Florence Nightingale felt as if she were experiencing "God" in the discipline of statistics, and she also used them in many practical ways for the betterment of society; U. S. President James A. Garfield saw the progress of all humankind in statistics, as well as the way to conduct the national census better than ever before; and baseball manager Billy Beane, of *Moneyball* fame, transformed his sport using "sabermetics"—a type of statistical analysis applied to baseball that allowed him to assemble a winning team at a fraction of the cost of some much richer franchises.

Some statisticians have thought they found the answers to truth and beauty in one statistic or statistical model or another: it was the arithmetic mean, or average, for Adolphe Quetelet; the normal or bell-shaped curve, for R.A. Fisher; probability itself for Simon Laplace and Thomas Bayes.

As the above suggests, a way to view many of the major statistical innovations is as a special type of "philosophy" about the world, especially human beings—since the statistical formulas and concepts purport to explain how to understand "truth," "beauty," "meaning," and other philosophical concepts, as well as how to achieve rational thinking, just outcomes in human relationships, epistemological credibility—the "Good Society," in short.

For some contemporary statisticians, their joy derives from the "power" of statistics—using statistics, for example, to win important court cases, or establish a correlation (or "link") between two or more important variables, to do "drug trials" on new chemical agents that can potentially help millions of people, or conduct studies that fill a void, add to a sparse body of knowledge, resolve a controversy, bridge a gap, or create new knowledge. By the "power" of statistics, in other words, I mean to indicate that statistics are commonly used today to determine or influence decisions that are responsible for our society to run, function, live up to the cultural and societal ideals expressed in our Constitution and Bill of Rights, and be a democracy. That is how essential statistics are to a modern society, and they are getting more so every day.

No Statistics Needed to Read this Book!

From the above you may think you need knowledge of statistical formulas or concepts to understand the readings in this book—but the good news is that you do not! All you need is an initial willingness to learn about some of the men and women who have invented statistics, applied them for the betterment of society, or wrote about them to positively influence others—plus a few hours of your time. Once you gain the motivation or inspiration to want to learn and use statistics, then you'll find that doing so will be a "piece of cake." It was for me, and I know it will be for you too!

The Ubiquity of Statistics and How They Are Used Today

As I hope you will appreciate from reading this book, statistics is one of the most important subjects you can learn today. Everywhere you turn, you hear about averages, probabilities, "links," projections, rates, percentages, sampling errors—to name only a few of the commonly reported statistics.

Statistics, in short, are near-ubiquitous in our society, to the point where they are even an integral part of our sports experience as fans, our weather reports and predictions, our retirement planning, our buying and selling behaviors . . . you name it, and statistics are probably involved in one way or another.

On television, the Internet, and radio, and in newspapers and magazines, statistics seem to be everywhere. They are used to lend credibility to arguments, describe phenomena, and make predictions. Graphs, charts, and tables are commonly used to visually depict "data"—the currency of our time for understanding the complex world in which we live.

"Big Data" has become the phrase of choice for journalists to describe the vast amount of bits and bytes residing on Fortune 500 servers around the world. And "data mining" has become the journalists' correlative phrase to describe the activity of extracting meaning from the Big Data, so companies like Google, Amazon, Facebook, Microsoft, IBM, and others can understand who their users are and how to appeal to them to buy consumer goods, among other things of value to corporate giants.

The government, too, has jumped into the business of mining Big Data full scale, even to the extent of gathering phone and Internet data on every citizen, purportedly for national security concerns, hoping that the "national net" of surveillance will result in the capture of the relatively few "terrorist sharks" who are hell bent on trying to devour America and its allies.

Because so many different types of organizations and individuals place so much value on data today, the field of "analytics" has emerged as a growing and valuable profession—a field simply defined as "the science of analysis," which involves the discovery and communication of meaningful patterns in data, mainly using knowledge of statistics and computers.

In sports, the science of "sabermetrics," mentioned above, has recently emerged to analyze statistical data and apply the results to baseball reporting and management, especially in order to evaluate and compare the performance of individual players.

In big cities, mayors now assemble "Geek squads" to help them understand their populations and devise policies.

Weather analysts depend on statistics to help them predict hurricanes, droughts, earthquakes, tsunamis, and other weather patterns.

And just about every company in the world that wants to grow its market share uses statistical techniques to analyze markets and create business plans that will help them achieve their short-range and long-term goals.

In academic journals, business books, government reports, watchdog organization alerts—and many other types of research-based publications—hundreds of thousands of studies are published every year around the world that present findings in the form of statistics. Many of them contain complex techniques that only advanced students know how to choose, execute, and interpret accurately, such as multivariate regression or path analysis, while other researchers use the simplest statistics to analyze their data and report the findings, such as frequencies, percentages, means, medians, modes, and standard deviations.

Whatever the statistic reported in a study, however, the point is that educated people today must know and understand basic statistics to be able to keep up with knowledge in their fields and make use of the information in an intelligent way.

Whether in medicine, sports, social science, biology, physics, chemistry, business, psychology, economics, entertainment, media—just about any discipline or field of activity today—the same applies: Statistical literacy is essential for being a well-informed citizen of a democratic, advanced technological society such as America. It is as essential as reading and writing have traditionally been, and for the same reasons.

What You Can Do with Statistics: How You Will Benefit by Being Statistically Literate

With knowledge of statistics, you'll be able to make better decisions about health and finance, to name only two important decision areas in life. You'll be able to assess whether what a newscaster or journalist is telling you makes any sense, and therefore whether you should act on the information. If you are politically inclined, then statistics can help you understand the bills, acts and laws that are proposed and passed or defeated, because many of them are based on statistical studies— quantitative "data," in a word. In general, you'll be able to understand life much better because so much of the information we receive these days, as mentioned above, is statistical—percentages, averages, correlations, and probabilities, among many other measures.

For all of these reasons and others you should want to know basic statistics—including how to understand them, compute them, and apply them, both in general during everyday life and in specific, limited-objective situations, such as doing your own market research study before launching a new product or service, deciding whether to buy or sell a house, analyzing your investment portfolio, among so many other possibilities.

Another general benefit of knowing statistics is that you will be able to "see the world" through "statistical eyes," which will not only provide you with new information you would not otherwise have, but also provide you with an alternative to the "expected reality" that an advertiser, politician or network executive—among so many other types of people—wants you to have. Statistical reality, in other words, will give you a way to *understand* the world you live in everyday in a way that is often contrary to the expectations of the pharmaceutical companies, the government, and the sports industries, among so many other groups and organizations that want you to see the world *their* way, based on the data or "results" that they choose to present to the public, which includes you, of course.

Using statistical thinking and know-how, you will also be able to *evaluate* what you read in the paper, about what a reported "average," "link" between variables, "prediction," or "rate of change" means from a seemingly countless stream of studies and reported statistics that flood our newspapers, magazines, television screens, radios, and Internet sites on a regular basis, everyday.

As you can see from the above, because statistics are so integral to and essential for our lives, *not* understanding statistics is a type of illiteracy that can and will prevent you from functioning well in our society, the same as having little or no ability to read or write will hold you back in what you try to do. Statistical literacy, in short, can help you to live a better, more profitable, safer, wealthier, successful, intelligent life. Knowledge of statistics is almost certain to make you a better consumer, decision-maker in business and personal life, citizen of society, fighter for personal and civil rights, and economic actor, among many other things that are important in your life.

For this reason, I believe you will agree that having facility with how to read, understand, interpret, and use statistics is a great asset, a great skill to have, and a great addition to your personal arsenal of cognitive and intellectual artillery.

About the Essays and Articles in this Book

Content and Sources: The readings in this book consist of essays originally written for this anthology, previously published articles, and excerpts from articles. The previously published articles and excerpts originally appeared in professional publications, popular periodicals, government documents, and websites, among other sources. At the end of every essay or article the source is stated, followed by references included in the article or essay.

Attribution: The name of the author of each essay originally written for this anthology appears after the title, and a brief biography of each author is presented at the end of the essay under the heading "Source"; and the name of each author of a reprinted article or essay whose name is designated is stated after the title. The readings that do not designate an author after the title include those by "institutional authors" (e.g., an article from the Bureau of Labor Statistics website without an author's name indicated) and essays created by the editorial staff from source materials about the topic or person in question (e.g., the essay titled "The Godliness of Statistics: Florence Nightingale" was originally prepared for this anthology by the editorial staff from existing documents).

Introductions to Readings: Each essay or article is preceded by a brief "introduction" that is intended to "set the stage," orient the reader, and highlight the importance or significance of the reading.

Organization: Each essay or article in this book conveys the joy of statistics, the power of statistics, or both, and could be read in any particular order. The readings after the "Prologue" and "Editor's Introduction," however, suggest a natural progression from the past to the present. For that reason, they are presented in an approximate chronological order, going from some of the founders of statistics to the current time, and spanning about three hundred years. As you will see, the historical path of statistics has gone from it being an inspirational discipline with a philosophical bent to a science and art that became adopted, praised, and institutionalized by society's most powerful leaders to an "applied" discipline that is ubiquitous in contemporary society and deeply affects virtually every aspect of life.

After the essays and articles are two sections, including *A Potpourri of Quotations about Statistics and Statisticians*, which presents a rich selection of quotations about statistics and statisticians expressed by philosophers, politicians, scientists, performing artists, and, of course, professional statisticians; and *Works Cited,* which contains all of the works cited in the anthology presented in alphabetical order by last name of the author.

Statistics have stimulated "religious" feelings in some people who feel they can reveal universal truths about the world and help us to achieve our highest goals, as may be seen in this essay about the nineteenth century statistical visionary, founder of modern nursing, and social reformer known to millions as "The Lady with the Lamp" and "The Queen of Nursing."

The "Godliness" of Statistics
*
Florence Nightingale

Most people know Florence Nightingale (1820-1910) as the legendary founder of modern nursing and a British social reformer who led the nurses caring for thousands of soldiers during the Crimean War, and thus helped save the British army from medical disaster.

Few people, however, know that Ms. Nightingale was a passionate statistician and pioneer in the visual depiction of statistical data. Statistics were so important to her, in fact, that her "hero as scientist"—the person she most admired in the world—was the Belgian statistician Adolphe Quetelet (Pearson, 1924), *the* most important founder of modern statistics according to many historians and contemporary data analysts. "What fascinated Miss Nightingale most about Quetelet was his *Essai de physique sociale* (first published in 1835), in which he showed the possibility of applying the statistical method to social dynamics, and deduced from such method various conclusions with regard to the physical and intellectual qualities of man" (Cook, 1913, p. 429).

Nightingale loved statistics for their practical uses, on the one hand, such as for decreasing preventable mortality among British soldiers; and she believed that administrators, politicians, and scientists could be successful only if they were guided by statistical knowledge. On the other hand, however, she was "a 'passionate statistician.' And the passion . . . was even a religious passion" (Cook, 1913, p. 435), because she viewed statistics as "heavenly" or "divine"—so divine, in fact, she *literally* saw God in the very nature of statistics—a vision few people, if any, have had, or stated so boldly and unabashedly in writing.

As Pearson (1924) wrote about her religious passion for statistics, "Her statistics were more than a study, they were indeed her religion . . . . She held that the universe—including human communities—was evolving in accordance with a divine plan; that it was man's business to endeavour to understand this plan and guide his actions in sympathy with it. But to understand God's thoughts, she held we must study statistics, for these are the measure of his purpose. Thus the study of statistics was for her a religious duty" (pp. 414-15).

According to Cook (1913), she held that "The laws of God were . . . discoverable by experience, research, and analysis; or, as she sometimes put it, the character of God was ascertainable, though His essence might remain a mystery. The laws of God were the laws of life, and these were ascertainable by careful, and especially by statistical, inquiry" (pp. 480-81).

Because of this viewpoint and passion, Nightingale often spoke of Adolphe Quetelet—the statistician mentioned above—in "religious" terms, and she viewed his book *Essai de physique sociale,* as "a religious work—a revelation of the "Will of God" (Cook, 1913, p. 480).

*

From reading the above, you may be wondering how Florence Nightingale could have been so inspired by statistics—which many people today, if not most, view as a method for "crunching numbers" in this age of computers, the Internet, and "Big Data." How could she see a "divine purpose" behind formulas designed to mechanically churn out "stats," such as arithmetic means, standard deviations, correlation coefficients, growth rates, and sampling errors?

By the time you are through reading this book, I hope you have some basis for understanding her feelings—even if you never personally reach such exalted heights—and develop both a deep appreciation for the role of statistics in our lives and a true motivation to want to learn and use statistics expertly.

References

Cook, Sir Edward. (1913). *The life of Florence Nightingale, in two volumes, vol. I (1820–1861)*. London: Macmillan and Co., Limited.

Pearson, Karl. (1924). *The Life, Letters and Labours of Francis Galton*, vol. 2. London: Cambridge University Press.

Quetelet, Adolphe. (1835). *Essai de physique sociale*. Paris: Bachelier.

The "average" is a dominant influence in basic statistics and probably the most frequently reported statistic in the mass media today, but for Adolphe Quetelet and many other early statisticians it was exalted as the key to understanding and defining what is best about the world, including individuals and nations, as explained in this essay.

The Magnificence of the Mean
*
Adolphe Quetelet

Lambert Adolphe Jacques Quetelet—known simply as Adolphe Quetelet (1796-1874)—was a Belgian statistician, mathematician, astronomer, and sociologist renowned for his application of statistics and probability theory to social phenomena. To many he is the "father" of modern statistics, who put the arithmetic mean, or average, on a pedestal that exalted it to a position of splendor on Earth and—it is safe to say without exaggeration—the universe as a whole.

Speaking of the "average man," for example, Quetelet (1835) wrote, "An individual who epitomized in himself, at a given time, all the qualities of the average man would represent at once all the greatness, beauty and goodness of that being" (cited in Porter, 1986, 102). Even more, he concluded that the mean, all by itself, can stand for the ideals of a society in every realm, be it politics, aesthetics, or morals.

According to Hankins, who analyzed Quetelet's work extensively in his doctoral dissertation, "It was one of Quetelet's repeated assertions that the average man was the type of perfection in beauty and goodness. Believing the race to be intellectually progressive, he held the average man to be the most perfect intellectually . . . to represent the type of the absolutely good" (1908, p. 63).

Quetelet's exaltation of the average may sound strange to you, a citizen of a hi-tech society where "individualism" and being "exceptional" and "unique" are extolled as virtues. But as Quetelet wrote, "The determination of the average man is not merely a matter of speculative curiosity; it may be of the most important service to the science of man and the social system. It ought necessarily to precede every other inquiry into social physics, since it is, as it were, the basis. The average man, indeed, is in a nation what the centre of gravity is in a body; it is by having that central point in view that we arrive at the apprehension of all the phenomena of equilibrium and motion" (1842, p. 96).

According to Meitzen, "Quetelet left no doubt that he was convinced of the presence of laws, capable of proof by calculation, which govern the life and actions of man and society" (1891, p. 85). But for Quetelet, "calculation" centered almost solely around the mean, which for him was ultimately important not just as a concept but also as a computational tool. As Porter wrote, Quetelet "almost never used mathematics in his empirical work on statistics. . . . The most sophisticated test Quetelet ever applied in published work to assess the reliability of his numbers was to divide the data at random into three groups and then compare their respective means" (Porter, 1986, p. 46).

Apart from any practical applications of the average, Hankins found "something fascinating," beautiful and ethereal in Quetelet's conceptualization of the average, as may be seen in the following passage:

> "It [the average] is a statistical conception of the universe possessing qualities of poetic and artistic beauty. Everything is to be viewed as varying about a normal state in a

manner to be accurately described by beautiful bell-shaped curves of perfect symmetry but of varying amplitude. Thus it is that the individual varies about his normal self; thus members of a group vary about their average; thus the men of a nation, viewed as individuals, vary about the average man of the nation; thus a nation varies about its normal state; and finally, inasmuch as the qualities of the average man change from time to time and place to place in obedience to general causes, to follow the course of the average man in the whole series of nations would give us, in Quetelet's view, the principles of a social physics, the true mechanics of human history. (1908, pp. 62-63)

Quetelet's argument for the power of the mean was so great that it influenced giants of social science such as Emile Durkheim, William Wundt, and Karl Marx, among many others. As Porter 1986, pp. 66-69) wrote, for example, Durkheim identified the "average type" with the normal, as opposed to the "pathological" in the *Rules of Sociological Method* (1895); "Wilhelm Wundt called for a natural history of human society based on the laws of statistics, arguing: 'It can be stated without exaggeration that *more psychology can be learned from statistical averages than from all philosophies, except Aristotle*'"; and Karl Marx, "explained how Quetelet's doctrine of the average man could be used to define a uniform standard of labor, and hence to furnish an exact and defensible interpretation of the labor theory of value." Marx also wrote in the *New York Daily Tribune* that [Quetelet's] *Physique Sociale* was "an excellent and learned work" (Beirne, 1993, p. 65).

Hankins aptly summarizes what Quetelet did for science, especially social science, by saying,

Quetelet's main contributions to social science were his demonstrations of and insistence upon the regularity and order in social phenomena and his formulation of a method for discovering this order. The exaltation of statistics into an exact instrument of observation was more uniquely his service than his contention that there are laws of human action and social life. . . . Quetelet's conception of the average man was based on the doctrine that in all that relates to social groups there will be found this variability about the group average. His statistical method therefore became a search for averages, for the limits of variation and for the manner in which this variation, under ordinary conditions, would occur. (1908, pp. 98-100)

References

Beirne, Piers. (1993). *Inventing criminality: Essays on the rise of 'Homo Criminalis.'* Albany, NY: State University of New York Press.

Hankins, Frank Hamilton. (1908). *Adolphe Quetelet as statistician.* New York: Columbia University, doctoral dissertation.

Quetelet, Adolphe. (1835). *Sur l'homme et let developpement de ses facultes ou essai de physique sociale.* Paris: Bachelier.

Quetelet, Adolphe. (1842). *A treatise on man and the development of his faculties.* Reprinted with an introduction by Solomon Diamond (1969). Gainesville: Scholars' Facsimiles and Reprints.

Porter, Theodore. (1986). *The rise of statistical thinking: 1820—1900.* Princeton, NJ: Princeton University Press.

Marx, Karl. (1867; 1978). *Capital*, volume one, edited by Robert C. Tucker. New York: Norton.

Meitzen, August (1891). *History, theory and technique of statistics*. Trans. by Roland Falkner. Philadelphia: American Academy of Political and Social Science.

Statisticians use probability theory, models, and values in many practical ways today, such as to test hypotheses and make predictions, but the major creators of probability in the eighteenth and nineteenth centuries viewed it "religiously" and felt it could eradicate flawed thinking and lead to the Good Society for all people, as this essay explains.

Probability: The Key to Reason and Religion
*
Thomas Bayes and Simon Laplace

Pierre-Simon, marquis de Laplace, generally known as Pierre-Simon Laplace (1749-1827) was a French mathematician and astronomer whose work was pivotal to the development of mathematical astronomy and statistics. One of the greatest scientists of all time, who is sometimes referred to as the "Newton of France," he summarized and extended the work of his predecessors in his five-volume *Mécanique Céleste* (*Celestial Mechanics*) (1799-1825).

In statistics Laplace mainly developed the interpretation of probability created by Reverend Thomas Bayes. His now-famous "theorem," which had no practical application in his lifetime, is widely used today by statisticians; and because of computers, it is routinely employed in the modeling of climate change, astrophysics and stock-market analysis (Davidson, 2010). He presented his theorem in *An Essay towards Solving a Problem in the Doctrine of Chances*, which was not published until after his death, when his friend, Richard Price, brought it to the attention of the British Royal Society in 1763. The theorem "concerned how we formulate probabilistic beliefs about the world when we encounter new data" (Silver, p. 241).

In his own time, however, Bayes hoped that the science of probability could be used to prove the existence of God, as well as in many practical, secular ways. As Richard Price (1763) wrote in the Introduction to *An Essay towards Solving a Problem in the Doctrine of Chances:*

> The purpose [of probability theory] I mean is to shew what reason we have for believing that there are in the constitution of things fixt laws according to which things happen, and that, therefore, the frame of the world must be the effect of the wisdom and power of an intelligent cause; and thus to confirm the argument taken from final causes for the existence of the Deity.

Laplace, like Bayes, also viewed probability as having a "higher purpose," despite his scientific rationality, and appears to have been absolutely "transported" by the study and application of probabilities. In his book titled *A Philosophical Essay on Probabilities* (1825), for example, he states, "It is *remarkable* that a science which began with the consideration of games of chance should have become the *most important object of human knowledge*" (p. 123, emphasis added).

His *Philosophical Essay*, in its totality, is a transcendent testament to the power of thinking in terms of probabilities. This type of thinking is needed, according to Laplace, to properly understand the world, overcome superstition, make rational decisions, enact laws, and uphold the three pillars of a democratic society: freedom, justice, and equality. As he says in his introduction,

> In this essay, I present here without the aid of analysis the principles and general results of this theory [of probabilities], applying them to the *most important questions of*

life, which are indeed for the most part *only problems of probability*. Strictly speaking it may even be said that nearly all our knowledge is problematical; and in the small number of things which we are able to know with certainty, even in the mathematical sciences themselves, *the principal means for ascertaining truth*—induction and analogy—*are based on probabilities*; so that *the entire system of human knowledge is connected with the theory set forth in this essay*" (pp. 1-2; emphasis added).

After presenting his fundamental assertions, Laplace discusses the power of probabilistic thinking for nearly 200 pages. With examples drawn from such diverse areas of life as gambling and courtroom trials, he convincingly argues that applying the theory of probabilities is the only way to have a rational, just, and free society, because probabilistic thinking serves as an "antidote" to the shortcomings of the human mind. It "makes us appreciate with exactitude," he says, "that which exact minds feel by a sort of instinct without being able ofttimes to give a reason for it. It leaves no arbitrariness in the choice of opinions and sides to be taken; and *by its use can always be determined the most advantageous choice*. Thereby *it supplements most happily the ignorance and the weakness of the human mind*" (p. 196; emphasis added)

More specifically Laplace concludes his philosophical essay by listing the specific qualities, benefits, and consequences of thinking in terms of probabilities, and leaves no doubt that such thinking is the path to the ideal human society, in these inspirational words:

If we consider the analytical methods to which this theory has given birth; the truth of the principles which serve as a basis; the fine and delicate logic which their employment in the solution of problems requires; the establishments of public utility which rest upon it; the extension which it has received and which it can still receive by its application to the most important questions of natural philosophy and the moral science; if we consider again that, even in the things which cannot be submitted to calculus, it gives the surest hints which can guide us in our judgments, and that it teaches us to avoid the illusions which ofttimes confuse us, then we shall see that *there is no science more worthy of our meditations, and that no more useful one could be incorporated in the system of public instruction* (p. 196; emphasis added).

References

Bayes, Thomas. (1763). *An Essay towards solving a problem in the doctrine of chances.* Philosophical Transactions:1683-1775. The Royal Society. http://www.jstor.org/stable/105741.

Davidson, Max. (2010). "Bill Bryson: Have faith, science can solve our problems": *Daily Telegraph* (26 September).

Laplace, Pierre-Simon. (1825; 1995). *Philosophical essay on probabilities.* Trans. By Andrew I. Dale. NY: Springer-Verlag.

Price, Richard. (1763). Introduction to *An essay towards solving a problem in the doctrine of chances,* by Thomas Bayes, Nov. 10. Philosophical Transactions: 1683-1775. The Royal Society. http://www.jstor.org/stable/105741.

Silver, Nate. (2012). *The signal and the noise: Why so many predictions fail—but some don't.* New York: The Penguin Press.

Statistics have inspired some of its great theorists and practitioners to "wax poetic" about seemingly mundane phenomena, such as national population changes—including the cousin of Charles Darwin discussed in this essay, who promoted the use of statistics in biology to make it scientific and developed several statistics in wide use today.

The "Charms" of Statistics
*
Sir Francis Galton

Sir Francis Galton (1822–1911) was an English scientist and a pioneer in the development of several statistical techniques widely used today, including the correlation coefficient, regression towards the mean, and standard deviation. Galton turned from exploration and meteorology to the study of heredity and eugenics—a term he coined that has taken on a dubious contemporary meaning—and established a system of classifying fingerprints still used today. His statistical work was carried on by his pupil Karl Pearson—one of the world's great statisticians—especially in developing the science of *biometrics* (discussed in the next essay). His best known book is, perhaps, *Inquiries into Human Faculty and Its Development* (1883). He was a cousin of Charles Darwin and knighted in 1909, two years before his death.

Galton had a droll, playful wit that he sometimes employed as a "weapon" against scientists with whom he disagreed, including the great French statistician Quetelet and his followers, who used the mean almost exclusively to statistically analyze phenomena. This may be seen, for example, in an essay Galton wrote titled the "The Charms of Statistics" (1894, p. 62)

> It is difficult to understand why statisticians commonly limit their inquiries to Averages, and do not revel in more comprehensive views. Their souls seem as dull to the charm of variety as that of the native of one of our flat English counties, whose retrospect of Switzerland was that, if its mountains could be thrown into its lakes, two nuisances would be got rid of at once. An Average is but a solitary fact, whereas if a single other *fact* be added to it, an entire Normal Scheme, which nearly corresponds to the observed one, starts potentially, into existence.

Galton goes on in the essay to praise statistics in words that reveal how inspired he was by the new science, on a level with Florence Nightingale, Thomas Bayes, and Simon Laplace, discussed above. He says,

> Some people hate the very name of statistics but I find them full of beauty and interest. Whenever they are not brutalised, but delicately handled by the higher methods, and are warily interpreted, their power of dealing with complicated phenomena is extraordinary. They are the only tools by which an opening can be cut through the formidable thicket of difficulties that bars the path of those who pursue the Science of man. (1894, pp. 62-63)

Galton was able to express his exuberance for statistics and describe worldly phenomena not only in ordinary prose, but also in "poetic" language, as if statistics were his "muse" and he were a "channel" for a higher power speaking through him. This poetic mode can clearly be seen in his description of population changes. Using almost pastoral language to discuss the "vital statistics of a nation," i.e., census data, he employs metaphors drawn from the military and nature to describe how individuals in a population change but the main features of the population remain the same.

> The vital statistics of a population are those of a vast army marching rank behind rank, across the treacherous tableland of life. Some of its members drop out of sight at every step, and a new rank is ever rising up to take the place vacated by the rank that preceded it, and which has already moved on. The population retains its peculiarities although the elements of which it is composed are never stationary, neither are the same individuals present at any two successive epochs. In these respects, a population may be compared to a cloud that seems to repose in calm upon a mountain plateau, while a gale of wind is blowing over it. The outline of the cloud remains unchanged, although its elements are in violent movement and in a condition of perpetual destruction and renewal. The well understood cause of such clouds is the deflection of a wind laden with invisible vapour, by means of the sloping flanks of the mountain, up to a level at which the atmosphere is much colder and rarer than below. Part of the invisible vapour, with which the wind was charged, becomes thereby condensed into the minute particles of water of which clouds are formed. After a while the process is reversed. The particles of cloud having been carried by the wind across the plateau, are swept down the other side of it again to a lower level, and during their descent they return into invisible vapour. Both in the cloud and in the population, there is on the one hand a continual supply and inrush of new individuals from the unseen; they remain a while as visible objects, and then disappear. The cloud and the population are composed of

elements that resemble each other in the brevity of their existence, while the general features of the cloud and of the population are alike in that they abide. (1889, pp. 164-65)

As can be seen, for Galton statistics had the power to help us understand the world deeply, as few other disciplines could, and he insisted they were needed to turn certain "theoretical" disciplines into sciences, including biology—an endeavor Galton pursued passionately with Karl Pearson in helping to create and publish the journal *Biometrika,* discussed in the next essay.

References

Galton, Francis. (1894). *Natural Inheritance.* New York and London: Macmillan and Co.

Galton, Francis. (1883). *Inquiries into Human Faculty and Its Development.* New York and London: Macmillan and Co.

Before quantitative methods were applied to the study of biological phenomena, biology was largely considered a "conjectural" discipline—including Charles Darwin's theory of evolution and natural selection—but this situation changed radically with the appearance of a new statistics-based journal whose mission was to turn biology into a science.

How Statistics Helped Make Biology a Science
*
Karl Pearson and *Biometrika*

Karl Pearson (1857-1936) was a brilliant English mathematician and statistician who specialized in mathematical biology or "biometry." He is often credited with establishing the discipline of mathematical statistics and was a major player in the early development of statistics as a serious scientific discipline in its own right. He founded the Department of Applied Statistics (now the Department of Statistical Science) at University College London in 1911—the first university statistics department in the world. Pearson created, along with Sir Francis Galton (discussed in the previous essay), the journal *Biometrika* and edited it for the first 35 years of its existence. Not only did he shape the journal, but he also contributed over two hundred pieces and inspired, more or less directly, most of the other contributions.

Pearson was passionate about publishing *Biometrika* mainly because Charles Darwin, the major influence on biological studies at the time, rarely used statistical methods in his studies or in formulating his theories. This led some critics to view Darwin's theories of evolution and natural selection as largely unscientific—a situation Pearson, along with Galton, sought to correct. They established *Biometrika* to address this charge and the shortcoming of biological studies in general. As Galton stated the mission of *Biometrika* in the very first edition, in an essay titled "Biometry," "This Journal is especially intended for those who are interested in the application to biology of the modern methods of statistics. Those methods deal comprehensively with entire species, and with entire groups of influences, just as if they were single entities, and express the relations between them in an equally compendious manner" (1901, p. 7).

Pearson introduced the first edition of *Biometrika* with an expansive "Editorial" in which he discussed "The Scope of *Biometrika*." Explaining the role of statistics in evolutionary theory, he began by stressing the role of *differences* or *variation* in individuals and species:

> The starting point of Darwin's theory of evolution is precisely the existence of those differences between individual members of a race or species . . . . The first condition necessary, in order that any process of Natural Selection may begin among a race, or species, is the existence of differences among its members; and the first step in an enquiry into the possible effect of a selective process upon any character of a race *must be an estimate of the frequency* with which individuals, exhibiting any given degree of abnormality with respect to that character, occur. The unit, with which such an enquiry must deal, is not an individual but a race, or a *statistically representative sample* of a race; and *the result must take the form of a numerical statement*, showing the relative frequency with which the various kinds of individuals composing the race occur. As it is with the fundamental phenomenon of variation, so it is with heredity and with selection. The statement that certain characters are selectively eliminated

from a race can be demonstrated *only by showing statistically* that the individuals which exhibit that character die earlier, or produce fewer offspring, than their fellows; while the phenomena of inheritance are only by slow degrees being rendered capable of expression in an intelligible form as numerical statements of the relation between parent and offspring, *based upon statistical examination of large series of cases*, are gradually accumulated.

These, and many other problems, *involve the collection of statistical data on a large scale*. That such data may be rendered intelligible to the mind, it is necessary to find some way of expressing them by a formula, the meaning of which can be readily understood, while its simplicity makes it easy to remember. *The recent development of statistical theory, dealing with biological data on the lines suggested by Mr. Francis Galton, has rendered it possible to deal with statistical data of very various kinds in a simple and intelligible way*, and the results already achieved permit the hope that simple formulae, capable of still wider application, may soon be found.

The number of biologists interested in these questions, and willing to undertake laborious statistical enquiries, is already considerable, and is increasing. It seems, therefore, that *a useful purpose will be served by a journal especially devoted to the publication of statistical data, and of papers dealing with statistical theory*. . . .

Whether we hold variation to be continuous or discontinuous in magnitude, to be slow or sudden in time, we recognise that *the problem of evolution is a problem in statistics, in the vital statistics of populations*. Whatever views we hold on selection, inheritance, or fertility, we must ultimately turn to the mathematics of large numbers, to the theory of mass-phenomena, to interpret safely our observations. *As we cannot follow the growth of nations without statistics of birth, death, duration of life, marriage and fertility, so it is impossible to follow the changes of any type of life without its vital statistics.* The evolutionist has to become in the widest sense of the words a registrar-general for all forms of life.

. . . every idea of Darwin—variation, natural selection, sexual selection, inheritance, prepotency, reversion—seems at once to fit itself to mathematical definition and to *demand statistical analysis*. Nor was the statistical conception itself entirely wanting in Darwin's work. The 'Cross and Self-Fertilisation of Plants' forms a splendid collection of statistical observations and experiments which offers many points of departure for further statistical research. . . .

We shall publish careful biometric observations, even if they be accompanied by only the most elementary statistical treatment; we shall look forward to our mathematical workers supplementing such fundamental observations by more elaborate statistical calculations. For this reason we shall not only print as copious observational and experimental data as possible, but endeavour to form a manuscript collection of such data available for further research. We hope that every number of *Biometrika* will *present statistical material ready for the mathematician to calculate and to reason upon*. All such investigations ancillary to data appearing in our pages we shall receive gladly and publish at the earliest opportunity." (1901, pp. 1-6; emphases added).

Although the passages above were written about a century ago, one can still feel Pearson and Galton's sense of joy and excitement in starting a publication devoted to establishing statistical analysis as an essential feature of biological studies. Their sense of promoting a cause and serving as "missionaries," one might say, for the developing science of statistics is almost palpable: One can feel their belief that statistics had the power to open new doors of perception and understanding that would further the study of biological phenomena as it advanced the progress of humankind.

References

Biometrika. (1901). Vol. 1, No. 1, Oct. Biometrika Trust. http://www.jstor.org/discover/10.2307/2331669?uid=3739560&uid=2134&uid=2&uid=70&uid=4&uid=3739256&sid=21102670168653

NOTE: Initial and later editions of *Biometrika* can be read at: http://www.jstor.org/action/showPublication?journalCode=biometrika

Charles Darwin was a "reluctant mathematician" but he inadvertently advanced the field of statistics when new techniques for assessing variation and testing hypotheses developed by Francis Galton, R. A. Fisher, and William Gosset were applied over several to analyze Darwin's plant experiments data.

Darwin: The Reluctant Mathematician
by
Julie Rehmeyer

For all his other talents, Charles Darwin wasn't much of a mathematician. In his autobiography, he writes that he studied math as a young man but also remembers that "it was repugnant to me." He dismissed complex mathematical arguments and wrote to a friend, "I have no faith in anything short of actual measurement and the Rule of Three," where the "Rule of Three" was an extremely simple mathematical calculation.

But history played a joke on the great biologist: It made him a contributor to the development of statistics.

It was the wildflower common toadflax that got the whole thing started. Darwin grew the plant for experiments, and he carefully cross-fertilized some flowers and self-fertilized others. When he grew the seeds, he found that the hybrids were bigger and stronger than the purebreds.

He was astonished. Although he had always suspected that inbreeding was bad for plants, he had never suspected it could have a significant effect within a single generation.

So he repeated the experiment with seven other kinds of plants, including corn. He had a clever, and at that time novel, idea. Since slight differences in soil or light or amount of water could affect the growth rates, he planted the seeds in pairs—one cross-pollinated seed and one self-pollinated seed in each pot. Then he let them grow and measured their heights.

Sure enough, on average, the hybrids were taller. Among his 30 corn plants, for example, the purebreds were only 84 percent as tall as the hybrids. But Darwin was savvy enough not to simply trust the average heights of so few plants. "I may premise," Darwin wrote, "that if we took by chance a dozen or score of men belonging to two nations and measured them, it would I presume be very rash to form any judgments from such small numbers on their average heights." Could it be, he wondered, that the height differences in the plants were just random variation?

Darwin noted, though, that men's heights vary a lot within a single country, whereas the heights of his plants didn't. His result might be more meaningful, but he wanted to be able to quantify how meaningful.

Doing that, however, required Darwin's hated mathematics.

So he turned to his cousin, Francis Galton, who just happened to be a leader in the emerging field of statistics. Galton had recently invented the standard deviation, a way of quantifying the amount of random variability in a set of numbers.

But Galton wasn't all that much use. He could calculate the standard deviation, but he couldn't use that number to tell Darwin how likely it was that the height difference wasn't just random. Furthermore, he was pretty sure it was too few plants to tell. "I doubt," he wrote, "after making many trials, whether it is possible to derive useful conclusions from these few observations. We ought to have measurement of at least fifty plants in each case, in order to be in a position to deduce fair results."

And there the matter rested, in frustrating uncertainty, for 40 years.

Resolving the impasse, it turned out, required some beer. The Guinness brewing company hired a young University of Oxford graduate, William Sealy Gosset, to develop statistical techniques to cheaply monitor the quality of its beer. The method Gosset developed was so powerful that it transformed statistics and continues to be a workhorse to this day.

Ironically, though, Gosset wasn't allowed to publish the method under his own name, because Guinness wanted to keep it a secret that statistics could help make better beer. But publish it he did, under the pseudonym "Student." The technique has hence become known as the "Student's t-test."

The Student's t-test did just what Galton didn't know how to do: Given the standard deviation Galton had calculated, it told how likely it was that the difference in the heights between the hybrids and the purebreds were just random. The answer? The chance was about one in 20. By statistical standards, that's significant, but barely so.

It took another 10 years and the intervention of another statistical genius for the next breakthrough on the problem. As a college student, Sir Ronald Aylmer Fisher learned about Gregor Mendel's work in genetics and Darwin's work in evolution, but the theory connecting the two hadn't yet been developed. Fisher set out to create the statistical foundation to make the connection possible. Darwin's experiment with hybrids was just the kind of problem Fisher needed to be able to solve.

He noticed something that Galton had missed: Galton had ignored Darwin's clever method of pairing the plants. He had calculated the standard deviation of the plants as a single, large group.

Fisher repeated the analysis but calculated the standard deviation of the difference in heights between the pairs of plants in each pot. Suddenly, instead of a one in 20 chance that the result didn't mean anything, he calculated about a one in 10,000 chance. In other words, it was nearly certain that the hybrids really did grow taller than the purebreds.

Fisher noted that the Student's t-test had one possible flaw: It assumed that the plant heights would vary in a predictable way (according to a normal distribution, to be precise). Just in case that assumption was wrong, he devised another way of analyzing the data and confirmed the result. "He was very clever in the way he did it," says Susan Holmes of Stanford University. Only in the 1980s did statisticians realize the full potential of Fisher's method and develop it into the subject of "exact testing."

Fisher's analysis was only possible because Darwin had designed his experiment so well. In fact, Fisher was often frustrated with the quality of other people's experiments. "To call in the statistician after the experiment is done," he said, "may be no more than asking him to perform a postmortem examination: he may be able to say what the experiment died of."

David Brillinger, a statistician at the University of California, Berkeley, says that Darwin's method of pairing is now common practice. "Darwin was a leader in a subfield of statistics called experimental design," he says. "He knew how to design a good experiment, but what to do with the numbers was something else."

Darwin himself came around eventually in his attitude toward mathematics. While he wrote in his autobiography of his youthful distaste for math, he also wrote that he wished he had learned the basic principles of math, "for men thus endowed seem to have an extra sense."

Source

Reprinted with Permission of Science News. Darwin: The Reluctant Mathematician." Julie Rehmeyer, *Science News*, February 11, 2009. http://www.sciencenews.org/view/generic/id/40740-/description/Darwin_The_reluctant_mathematician_

The U. S. Census relies on statistics as the way to analyze and summarize the demographic characteristics of our population and keep track of changes—as described in this brief excerpt from the introduction to Statistical View of the United States *written by the Superintendent of the U. S. Census in 1854.*

Statistical View of the United States
by
D. B. Debow

Among the Greeks and Romans inquiries in regard to population were often pressed to a considerable extent, yet the science of statistics, as now understood, may be said to belong altogether to the present age. . . .

The importance of correct information regarding the age, sex, condition, occupation and numbers of a people, their moral and social state, their education and industry, is now universally recognized among the enlightened of all civilized nations. Where this information can be had for periods running back very far, and for many countries, it furnishes the material for contrasts and comparisons the most instructive, and for deducing the soundest rules in the administration of Government, or in promoting the general welfare of society.

Statistics are far from being the barren array of figures ingeniously and laboriously combined into columns and tables, which many persons are apt to suppose them. They constitute rather the ledger of a nation, in which, like the merchant in his books, the citizen can read, at one view, all of the results of a year or of a period of years, as compared with other periods, and deduce the profit or the loss which has been made, in morals, education, wealth or power.

Source

Debow, D. B. (1854). *Statistical view of the United States, embracing its territory, population—white, free colored, and slave-moral and social condition, industry, property, and revenue; the detailed statistics of cities, towns, and counties; being a compendium of the seventh census, to which are added the results of every previous census, beginning with 1790, in comparative tables, with explanatory and illustrative notes, based upon the schedules and other official sources of information.* Washington: Beverly Tucker, Senate printer. The above excerpt is from Introductory Remarks, page 9.

Our twentieth president heralded statistics as a civilizing and progressive force in society to a degree unparalleled by any national leader, and for specific reasons that prominently include the census, as can be seen in this extended excerpt from one of his addresses to Congress when he was a member of the House of Representatives.

The Importance of Statistics to the United States Census and the World
by
President James A. Garfield

James A. Garfield was President of the United States from only March 4 to September 19, 1881, when he was assassinated. Before becoming president, he served in the House of Representatives for nine consecutive terms, from 1863-1881. During his terms in office, he gave various speeches on the subject of statistics—which was unique for a congressional member to do—that reflected his deep interest in both political science and the scientific thought of the age. In 1869, he introduced a resolution to examine the need for legislation regarding the ninth census to be taken the following year. In that speech (*Congressional Globe,* 1869, pp. 178-79), he succinctly summarized the importance of statistics, not only for the U.S. Census but for the good of all humankind, in the following words:

> The modern census is so closely related to the science of statistics that no general discussion of it is possible without considering the principles on which statistical science rests and the objects which it proposes to reach.

The science of statistics is of recent date, and, like many of its sister sciences, owes its origin to the best and freest impulses of modern civilization. The enumerations of inhabitants and the appraisements of property made by some of the nations of antiquity were practical means employed sometimes to distribute political power, but more frequently to adjust the burdens of war, but no attempt was made among them to classify the facts obtained so as to make them the basis of scientific induction. The thought of studying these facts to ascertain the wants of society had not then dawned upon the human mind, and, of course, there was not a science of statistics in this modern sense.

It is never easy to fix the precise date of the birth of any science, but we may safely say that statistics did not enter its scientific phase before 1749, when it received from Professor Achenwall, of Göttingen, not only its name, but the first comprehensive statement of its principles. Without pausing to trace the stages of its growth, some of the results of the cultivation of statistics in the spirit and methods of science may be stated as germane to this discussion:

1. It has developed the truth that society is an organism, whose elements and forces conform to laws as constant and pervasive as those which govern the material universe; and that the study of these laws will enable man to ameliorate his condition, to emancipate himself from the cruel dominion of superstition, and from countless evils which were once thought beyond his control, and will make him the master rather than the slave of nature....

The scientific spirit has cast out the demons, and presented us with nature clothed and in her right mind and living under the reign of law. It has given us, for the sorceries of the alchemist, the beautiful laws of chemistry; for the dreams of the astrologer, the sublime truths of astronomy; for the wild visions of cosmogony, the monumental records of geology; for the anarchy of diabolism, the laws of God. But more stubborn still has been the resistance against every attempt to assert the reign of law in the realm of society. In that struggle, *statistics has been the handmaid of science, and has poured a flood of light upon the dark questions of famine and pestilence, ignorance and crime, disease and death.* . . . We are beginning to acknowledge that —

"The fault, dear Brutus, is not in our stars,

But in ourselves, that we are underlings."

Governments are only beginning to recognize these truths. . . .

2. The development of statistics are [*sic*] causing history to be rewritten. Till recently the historian studied nations in the aggregate, and gave us only the story of princes, dynasties, sieges, and battles. Of the people themselves—the great social body with life, growth, sources, elements, and laws of its own—he told us nothing. Now statistical inquiry leads him into the hovels, homes, workshops, mines, fields, prisons, hospitals, and all places where human nature displays its weakness and its strength. In these explorations he discovers the seeds of national growth and decay, and thus becomes the prophet of his generation. . . .

3. *Statistical science is indispensable to modern statesmanship.* In legislation as in physical science it is

beginning to be understood that we can control terrestrial forces only by obeying their laws. The legislator must formulate in his statutes not only the national will, but also those *great laws of social life revealed by statistics*. He must study society rather than black-letter learning. He must learn the truth "that society usually prepares the crime, and the criminal is only the instrument that accomplishes it;" that statesmanship consists rather in removing causes than in punishing or evading results.

Light is itself a great corrective. A thousand wrongs and abuses that grow in the darkness disappear like owls and bats before the light of day. . . .

I know of no writer who has exhibited the importance of this science to statesmanship so fully and so ably as Sir George Cornwall Lewis, in his treatise *On the Methods of Observation and Reasoning on Politics*.

After showing that politics is now taking its place among the sciences, and as a science its superstructure rests on observed and classified facts, he says of the registration of political facts, *which consists of history and statistics*, that "it may be considered as the entrance and propylaea to politics. It furnishes the materials upon which the artificer operates, which he hews into shape and builds up into a symmetrical structure."

In a subsequent chapter, he portrays the importance of statistics to the practical statesman in this strong and lucid language:

"He can hardly take a single safe step without consulting them. Whether he be framing a plan of finance, or considering the operation of an existing tax, or following the variations of trade, or studying the public health, or examining the effects of a criminal law, *his conclusions ought to be guided by statistical data.*" — Vol. i, p. 134.

Napoleon, with that wonderful vision vouchsafed to genius, saw the importance of this science when he said:

"Statistics is the budget of things; and without a budget there is no public safety."

We may not, perhaps, go as far as Goethe did, and declare that "figures govern the world;" but we can fully agree with him that "they show how it is governed."

Baron Quetelet, of Belgium, one of the ripest scholars and profoundest students of statistical science, concludes his latest chapter of scientific results in these words:

"One of the principal results of civilization is to reduce more and more the limits within which the different elements of society fluctuate. The more intelligence increases the more these limits are reduced, and the nearer we approach the beautiful and the good. . . . and the more we advance, the less we shall have need to fear those great political convulsions and wars and their attendant results, which are the scourges of mankind."

It should be added that the growing importance of political science, as well as its recent origin, is exhibited in the fact that *nearly every modern nation has established within the last half century a bureau of general statistics for the uses of statesmanship and science.* In the thirty states of Europe they are now assiduously cultivating the science. Not one of their central bureaus was fully organized before the year 1800.

The chief instrument of American statistics is the census, which should accomplish a two-fold object. It should serve the country by making a full and accurate exhibit of the elements of national life and strength, and it should serve the science of statistics by so exhibiting general results that they may be compared with similar data obtained by other nations.

Source

Garfield, James A. (1869). *The Congressional Globe.* UNT Digital Library. http://digital.library.unt.edu/ark:/67531/metadc30-883/m1/550/sizes/?q=achenwall

Reference

Ridpath, John Clark. (1881). *The life and work of James A. Garfield.* Cincinnati, OH: Jones Brothers. http://archive.org/-stream/lifeandworkjame02ridpgoog/lifeandworkjame02ridpgoog-_djvu.txt

The U. S. Bureau of Labor Statistics is a general-purpose statistical agency that gathers, analyzes, and distributes information on many aspects of national life and contributes to policy development necessary for the government to function, as this essay describes, which recounts the agency's first one hundred years.

The First Hundred Years of the Bureau of Labor Statistics (1884-1984)
by
Joseph P. Goldberg and William T. Moye

The mission of the Bureau of Labor Statistics since its founding 100 years ago has been to collect information on economic and social conditions and, in the words of Carroll Wright, the first Commissioner, through "fearless publication of the results," to let the people assess the facts and act on them. It was the belief of its founders that dissemination of the facts would lead to improvement of the life of the people.

On the occasion of the Bureau's centennial, Janet L. Norwood, the tenth Commissioner, summed up the Bureau's past—and continuing—role: "The Bureau stands for—

—Commitment to objectivity and fairness in all of its data gathering and interpretive and analytical work;
—Insistence on candor at all times;
—Protection of confidentiality;
—Pursuit of improvements;
—Willingness to change; and
—Maintenance of consistency in the highest standards of performance."

These principles, Norwood stressed, must be steadfastly applied in monitoring "our programs to ensure that they remain accurate, objective, and relevant. We must modernize our statistical techniques because a statistical agency that does not constantly move ahead in the use of new techniques quickly moves backward."

As an institution, the Bureau has evolved from the original and sole labor agency in the Federal Government, with a broad fact-finding scope, to one among many specialized labor agencies. Serving as a quasi-Department of Labor during its first two decades, it was called upon to study and report on issues such as the violent strikes and lockouts of the period and the harsh conditions of employment for women and children. Today, the Bureau is a general-purpose statistical agency, gathering, analyzing, and distributing information broadly applicable to labor economics and labor conditions.

While the focus and perspectives of Bureau studies have changed over the years, most areas of investigation have remained germane—the course of wages and prices, the state of industrial relations, problems of unemployment and the effects of technological and demographic change, and safety and health conditions in the workplace. Some areas of study, such as child labor, have been rendered unnecessary by legislation. In others, newer, specialized agencies have taken over the work the Bureau began.

The Bureau's role has been to provide data and analyses that contribute to the development of policy without crossing the line into policy formulation, but the line is a fine one....

Professional integrity is essential to a government agency which provides information for public and private policy needs, and the Bureau's institutional probity has been a constant concern of the Commissioners and their staffs. Over the years, the Bureau's objectivity has been affirmed and reaffirmed upon review of its work by congressional committees, Presidential commissions, and professional associations of economists and statisticians. All noted areas needing improvement, but none found reason to question the independence and integrity of the Bureau.

For the first half-century of its existence, the Bureau's appropriations changed only when special funding was provided for particular programs, such as the woman and child study of 1907-09, and development of a cost-of-living index during World War I. The emergency demands of the depression of the 1930's and the accompanying social legislation also led Congress to increase appropriations to expand and improve the Bureau's statistics. And, similarly, World War II needs generated increased resources and programs.

After the war, the climate was vastly different. Government policy concerns required data produced on a frequent and regular basis. The Employment Act of 1946, which established the congressional Joint Economic Committee and the Council of Economic Advisers, epitomized the new conditions. As government social and economic policies developed and expanded, legislation frequently incorporated Bureau statistics as escalators or other administrative devices. There was now a regular demand for new and improved statistics, with support for resources to make them available. While increases in resources have not always been forthcoming, and programs have been cut on occasion to make room for new and expanded series, the postwar trend has been one of provision of funds for such expansion and improvements.

Bureau programs have changed to meet changing conditions. Ongoing statistical series such as the Consumer Price Index have been adjusted periodically to assure that concepts and coverage reflect altered societal patterns. Along with regular planned revisions, the Bureau has made interim revisions, as in the case of the treatment of the homeownership component in the CPI. New series, including the Employment Cost Index and the multifactor productivity indexes, have been developed.

Meeting these vastly increased requirements has been made possible through the development of sophisticated statistical techniques of sampling and the computerization of statistical operations. Bureau personnel now include mathematical statisticians, computer programmers, and computer systems analysts as well as economists and clerical staff.

In addition, close coordination with other Federal agencies and with the States, evolving from Wright's early efforts, has improved the quality of the data and efficiency in collection and processing.

Bureau respondents generally have given their full cooperation because of the assurance of confidentiality for reported information, a guarantee which has been assiduously enforced. In its communication with the public, the Bureau has emphasized frankness regarding limitations of the data and the provision of detailed information on concepts and methods. There has been a constant striving to improve the timeliness, regularity, and accuracy of the data and their public presentation.

While well established, the principles have needed regular reiteration, particularly during unsettled times. There have been many occasions when the messenger has been buffeted by the storms of rapid economic and social change. This has been especially true when the Bureau's data have been used in implementing and monitoring policy, as in the wartime use of its cost-of-living index for wage stabilization. On other occasions, Bureau staff efforts to explain technical limitations have collided with policymakers' unqualified use of the data. In such circumstances, the Bureau has been sustained by the widespread recognition that its nonpartisanship and objectivity must be assured and protected. Congress, successive Secretaries of Labor, the Bureau's labor and business advisory groups, the professional associations, and the press have supported the independence and impartiality of statistical research in government agencies.

The roots of this independence and professionalism are deep and strong. The tradition of impartiality has been underwritten by both Democratic and Republican administrations over the century of the Bureau's existence, during which Commissioners have been selected for their technical competence without regard to partisan considerations.

The Bureau faces great challenges in the years ahead as the phenomena it measures grow in complexity in the dynamic economy of the United States. It will require openness to new methods and techniques and adherence to the standards already set to carry out its mission during the next century.

Source

Goldberg, Joseph P. & Moye, William T. (1985). *The first hundred years of the Bureau of Labor Statistics* (1884-1984). Washington, DC: U. S. Government Printing Office. The excerpt above is Chapter IX: History as Prologue: The Continuing Mission. Online at http://www.bls.gov/opub/blsfirsthundredyears/-100_years_of_bls.pdf).

Social statistics have long played a crucial role in many aspects of our military and government-related activities, as discussed in the following brief excerpts from a Presidential Address delivered at the Eightieth Annual Meeting of the American Statistical Association, in Richmond, Virginia, December 1918.

Statistics and Government
by
Wesley C. Mitchell

Since the American Statistical Association was founded in 1839 no year has brought such stirring changes in American statistics as the year now closing. The war forced a rapid expansion in the scope of federal statistics and the creation of new statistical agencies. What is more significant, the war led to the use of statistics, not only as a record of what had happened, but also as a vital factor in planning what should be done. The war also brought an unprecedentedly large number of statisticians into government employ. Probably there are few professional societies which have had so considerable a proportion of their membership engaged in war work as this Association.

Tonight we feel a just pride in the service our fellow members have rendered. We cherish the hope that what they have helped to accomplish during the war toward the guidance of public policy by quantitative knowledge of social fact may not be lost in the period of reconstruction through which we are passing, and in the indefinite period of peace upon which we are about to enter. To forward that hope the Association may seek a more active share in the work of federal statistics in the future than it has ever taken in the past. . . .

In speaking next of our hopes for the future, I am speaking merely as one member of the American Statistical Association. Yet I believe that most members of our Association believe that the social sciences in general and social statistics in particular have a great service to render to government and through government to mankind.

The episode in statistical organization which I have sketched, the effect of the war upon our attitude toward the use of facts for the guidance of policy, links the present stage of civilization with man's savage past. . . .

The social sciences, however, cover an immense field, and it is not probable that we shall encounter failure or success in all its parts. The parts where effort seems most promising just now are the parts in which this Association is particularly interested. Measurement is one of the outstanding characteristics of science at large, whether in the field of inorganic matter or life processes. Social statistics, which is concerned with the measurement of social phenomena, has many of the progressive features of the physical sciences. It shows forthright progress in knowledge of fact, in technique of analysis, and in refinement of results. It is amenable to mathematical formulation. It is capable of forecasting group phenomena. It is objective. A statistician is usually either right or wrong, and his successors can demonstrate which. Statisticians are not continually beginning their science all over again by developing new viewpoints. Where one investigator stops, the next investigator begins with larger collections of data, with extensions into fresh fields, or with more powerful methods of analysis. In all these respects, the position and prospects of social statistics are more like the position and prospects of the natural sciences than like those of the social sciences.

Above all, social statistics even in its present state is directly applicable over a wide range in the management of practical affairs, particularly the affairs of government. And this practical value of statistics is readily demonstrable even to a busy executive. Once secure a quantitative statement of the crucial elements in an official's problem, draw it up in concise form, illuminate the tables with a chart or two, bind the memorandum in an attractive cover tied with a neat bow-knot, and it is the exceptional man who will reject your aid. Thereafter your trouble will be not to get your statistics used, but to meet the continual calls for more figures, and to prevent your convert from taking your estimates more literally than you take them yourself.

We may well cherish high hopes for the immediate future of social statistics. In contributing toward a quantitative knowledge of social facts, in putting this knowledge at the disposal of responsible officials, we are contributing a crucially important part toward achieving the gravest task that confronts mankind today—the task of developing a method by which we may make cumulative progress in social organization.

Source

Mitchell, Wesley C. (1919). Statistics and government. *American Statistical Association, New Series, No. 125*, 223-235.

Statistics are often allowed as evidence in civil and criminal court cases today, but such evidence was rarely allowed in courtrooms as late as the 1950s—a situation that radically changed when the U. S. Supreme Court cited statistical findings from studies on children to bolster its constitutional ruling in a case many consider to be the most important of the twentieth century.

How Statistics Helped Desegregate America's Public School System
*
Kenneth Clark and the U. S. Supreme Court

Until 1954, America's public schools were segregated on the basis of "race" in 21 states, i.e., black children went to all black schools and white children went to all white schools. On May 17, 1954, however, the Supreme Court rendered what is arguably its most important decision of the twentieth century: *Brown v. the Board of Education, 347 U.S. 483.*

In delivering the opinion of the Court, Chief Justice Earl Warren said, "We conclude that in the field of public education the doctrine of 'separate but equal' has no place. Separate educational facilities are inherently unequal."

In reaching this decision, which overturned the ruling of "separate but equal" rendered in the case of "Plessey versus Ferguson," the justices knew their decision would be severely criticized on constitutional grounds alone. As a result, they gave great weight to seven social science and psychological studies—which they referred to as "modern authority" and cited in the famous Footnote 11 of their landmark ruling—to lend scientific credibility to their conclusion that racially segregated schools caused black students to feel inferior to white children, and were thus inherently unequal.

Among the seven studies, the one that carried the most weight was the Dolls Test, which is replete with statistics and was conducted many times over the years by the social psychologist Kenneth Clark.

So important was the Dolls Test, that *The New York Times* obituary of Kenneth Clark, who died on May 1, 2005, at 90 years of age, had the headline: "Kenneth Clark, academic who helped to desegregate U.S. schools," and went on to state:

> Kenneth Clark, the psychologist and educator whose 1950 report showing the deleterious effect of school segregation influenced the U.S. Supreme Court to hold school segregation to be unconstitutional, died on Sunday in Hastings-on-Hudson, New York. . . . It was his research with black schoolchildren that became a pillar of *Brown v. the Board of Education*, the 1954 Supreme Court decision that toppled the "separate but equal" doctrine of racial segregation that then prevailed in 21 states. (Severo, 2005).

What, exactly, was the Dolls Test? What kind of statistics were calculated that proved to be so powerful in helping the Supreme Court reach the momentous decision to desegregate the public schools in America? The answer to the latter question may surprise you: The main statistics were the most elementary kind—frequencies and percentages—sometimes accompanied by a "critical ratio" of the difference between the percentages or the result of a "significance test," done to determine if the percentages for the samples differed "beyond chance." It is the very simplicity of these statistics that truly proves the power statistics can have when presented by a respected authority in the right context at the right time.

The Dolls Test was originally designed during the 1940s by Kenneth Clark and his wife, Mamie Phipps Clark, as a way to measure the psychological effects of segregation on black children. Basically, the test involved showing black children from different parts of the country (either racially segregated or integrated) both black and white dolls, and then asking the children what dolls they identified with, what dolls they liked the most, or what dolls they thought looked most like them, among other related questions.

In the original Dolls Test, 253 "Negro children," as the Clarks referred to them, composed the total sample, with 134 in the "southern group"—tested in segregated nursery schools in Arkansas—and 119 in the "northern group"—tested in racially mixed nursery schools in Massachusetts. The children were between three and seven years old, including 116 boys and 137 girls. The Clarks also classified the children by their skin tone, including 46 who were "light" or "practically white," 128 who were "medium" or "light brown to dark brown," and 79 who were "dark" or "dark brown to black."

In their analyses the Clarks cross-classified the children according to the above three variables: their geographic location, sex, and skin tone. For example, they classified the number of boys who were "light" and three years old; the number of girls who were "dark" and seven years old, etc.

When the Clarks showed the white and black dolls to the children, although all the children were black, most (a) identified themselves with the white dolls, (b) attributed more positive characteristics to the white dolls, and (c) thought they looked more like the white dolls than the black dolls. The results varied to different degrees, however, according to the children's gender, age, skin tone, and residential area (north or south). The results were most striking for the northern group, who lived in areas with many white people, whereas the southern group knew only black people and apparently had a firmer sense of their own self-identity.

To organize and record the children's answers, the Clarks used printed data sheets, and also wrote down their general observations. In statistical table after table, they showed the frequency, or number, of children with different age, color, gender, and regional characteristics, along with how they answered a question about the dolls. In some tables they also showed percentages to go along with the frequencies; and in certain instances, as mentioned, they reported that they conducted a "significance test" to see if the results obtained *between* different groups, e.g., north and south, boys and girls, were statistically "significant," or different beyond what one would expect by chance.

The children's answers were then tallied, percentages were calculated, and sometimes, as mentioned, a critical ratio or result of a significance test accompanied the frequencies and percentages. For example, a typical statement from one of their papers, referring to the two samples of black children, reads as follows:

> Fifty-nine percent of these children indicated that the colored doll "looks bad," while only 17 percent stated that the white doll "looks bad" (critical ratio 10.9). That this preference and negation in some way involve skin color is indicated by the results for request 4. Only 38 percent of the children thought that the brown doll was a "nice color," while 60 percent of them thought that the white doll was a "nice color" (critical ratio 5.0).
>
> The importance of these results for an understanding of the origin and development of racial concepts and attitudes in Negro children cannot be minimized. (1947, p. 175).

The statistical findings from the Dolls Tests played a significant role, as mentioned, in providing a rationale for the Supreme Court to rule, in part, as follows, in its landmark decision:

> Segregation of white and colored children in public schools has a detrimental effect upon the colored children. The impact is greater when it has the sanction of the law, for the policy of separating the races is usually interpreted as denoting the inferiority of the Negro group. A sense of inferiority affects the motivation of a child to learn. Segregation with the sanction of law, therefore, has a tendency to [retard] the educational and mental development of Negro children and to deprive them of some of the benefits they would receive in a racially integrated school system.

The Court thus held that segregation was a violation of the Fourteenth Amendment Equal Protection Clause in the Constitution and declared segregation in public schools to be unconstitutional.

References

NOTE: The following are the citations for the seven studies cited by the Supreme Court, including the Dolls Study by Kenneth Clark, as presented in footnote 11:

Clark, K. B. Effect of Prejudice and Discrimination on Personality Development (Midcentury White House Conference on Children and Youth, 1950); Witmer and Kotinsky, Personality in the Making (1952), c. VI; Deutscher and Chein, The Psychological Effects of Enforced Segregation: A Survey of Social Science Opinion, 26 J. Psychol. 259-287 (1948); Chein, What are the Psychological Effects of Segregation Under Conditions of Equal

Facilities?, 3 Int. J. Opinion and Attitude Res. 229 (1949); Brameld, Educational Costs, in Discrimination and National Welfare (MacIver, ed., (1949), 44-48; Frazier, The Negro in the United States (1949), 674-681. And see generally Myrdal, An American Dilemma (1944).

Clark, Kenneth B. & Clark, Mamie, P. (1947). Racial identification and preference in Negro children. Condensed by the authors from an unpublished study. http://i2.cdn.turner.com/cnn-/2010/images/05/13/doll.study.1947.pdf

Severo, Richard. (2005). Obituary of Kenneth Clark. *New York Times*, Tuesday, May 3. http://www.nytimes.com/2005/05-/02/nyregion/02clark.html?_r=0

Many statistical studies published over several decades showed a strong correlation between smoking, cancer, and mortality, which led the U. S. Surgeon General to issue a landmark report in 1964 declaring cigarettes, pipes, and cigars unhealthy—described in the excerpts below—and caused tens of millions of Americans to suddenly quit smoking.

The 1964 Surgeon General's Report on Smoking

The Effects of Smoking: Principal Findings

Cigarette smoking is associated with a 70 percent increase in the age-specific death rates of males, and to a lesser extent with increased death rates of females. The total number of excess deaths causally related to cigarette smoking in the U.S. population cannot be accurately estimated. In view of the continuing and mounting evidence from many sources, it is the judgment of the Committee [the Surgeon General's Advisory Committee on Smoking and Health] that cigarette smoking contributes substantially to mortality from certain specific diseases and to the overall death rate.

Lung Cancer

Cigarette smoking is causally related to lung cancer in men; the magnitude of the effect of cigarette smoking far outweighs all other factors. The data for women, though less extensive, point in the same direction. The risk of developing lung cancer increases with duration of smoking and the number of cigarettes smoked per day, and is diminished by discontinuing smoking. In comparison with non-smokers, average male smokers of cigarettes have approximately a 9- to 10-fold risk of developing lung cancer and heavy smokers at least a 20-fold risk. The risk of developing cancer of the lung for the combined group of pipe smokers, cigar smokers, and pipe and cigar smokers is greater than for non-smokers, but much less than for cigarette smokers. Cigarette smoking is much more important than occupational exposures in the causation of lung cancer in the general population. . . .

Mortality

The death rate for smokers of cigarettes only, who were smoking at the time of entry into the particular prospective study, is about 70 percent higher than that for non-smokers. The death rates increase with the amount smoked. For groups of men smoking less than 10, 10-19, 20-39, and 40 cigarettes and over per day, respectively, the death rates are about 40 percent, 70 percent, 90 percent, and 120 percent higher than for non-smokers. The ratio of the death rates of smokers to non-smokers is highest at the earlier ages (under 50) represented in these studies, and declines with increasing age. The same effect appears to hold for the ratio of the death rate of heavy smokers to light smokers. In the studies that provided this information, the mortality ratio of cigarette smokers to non-smokers was substantially higher for men who started to smoke under age 20 than for men who started after age 20. The mortality ratio was increased as the number of years of smoking increased. In two studies which recorded the degree of inhalation, the mortality ratio for a given amount of smoking was greater for inhalers than non-inhalers. Cigarette smokers who had stopped smoking prior to enrollment in the study had mortality ratios about 1.4 as against 1.7 for current cigarette smokers. The mortality ratio of ex-cigarette smokers increased with the number of years of smoking and was higher for those who stopped after age 55 than for those who stopped at an earlier age (Chapter 8, p. 93). The biases from non-response and from errors of measurement that are difficult to avoid in mass studies may have resulted in some over-estimation of the true mortality ratios for the complete populations. In our judgment, however, such biases can account for only a part of the elevation in mortality ratios found for cigarette smokers (Chapter 8, p. 96).

Death rates of cigar smokers are about the same as those of non-smokers for men smoking less than five cigars daily. For men smoking five or more cigars daily, death rates were slightly higher (9 percent to 27 percent) than for non-smokers in the four studies that gave this information. There is some indication that this higher death rate occurs primarily in men who have been smoking for more than 30 years and in men who stated that they inhaled smoke to some degree. . . .

The mortality ratio of cigarette smokers was particularly high for a number of diseases. There is a further group of diseases, including some of the most important chronic diseases, for which the mortality ratio for cigarette smokers lay between 1.2 and 2.0. The explanation of moderate elevations in mortality ratios in this large group of causes is not clear. Part may be due to the sources of bias previously mentioned or some constitutional and genetic difference between cigarette smokers and non-smokers. There is also the possibility that cigarette smoking has some general debilitating effect, although no medical evidence that clearly supports this hypothesis can be cited (Chapter 8, p. 105). . . .

Coronary artery disease is the chief contributor to excess number of deaths of cigarette smokers over non-smokers, with lung cancer uniformly in second place (Chapter 8, p. 108). For cigar and pipe smokers combined, there was a suggestion of high mortality ratios for cancers of the mouth, esophagus, larynx and lung, and for stomach and duodenal ulcers. These ratios are, however, based on small numbers of deaths (Chapter 8, p. 107).

*

[NOTE: A year after the publication of the Surgeon General's Report, the U. S. Congress passed the Federal Cigarette Labeling and Advertising Act of 1965, which required warning labels on cigarette packs, and four years later the Public Health Cigarette Smoking Act of 1969 required similar warnings on print advertising for cigarettes.]

Source

Smoking and health: a report of the advisory committee to the Surgeon General of the Public Health Services. (1964). Released by U.S. Surgeon General Luther Terry. Online at: http://profiles.nlm.nih.gov/ps/access/NNBBMT.pdf. The excerpts above are from pp. 31, 35-37.

References

The major statistical studies referred to in the Surgeon General's Report include the following, as they appear in the Report:

Best E., Josie, G., & Walker, C. A Canadian study of mortality in relation to smoking habits, a preliminary report. *Canad J Pub Health.* 1961; 52:99-106. (A mortality analysis of 78,000 Canadian pensioners.)

Doll, R. & Hill, A. Lung cancer and other causes of death in relation to smoking. *Brit Med J.* 1956; 2:1071-1081. (A mortality analysis commencing in 1951 of 34,000 British physicians.)

Dunn, J. Jr., Linden, G. & Breslow, L. Lung cancer mortality experience of men in certain occupations in California. *Am J Pub Health.* 1960; 50:1475-1487. (A mortality analysis of 67,000 male subjects.)

Hammond, E. & Horn, D. Smoking and death rates—reported on forty-four months of follow-up on 187,783 men. *JAMA.* 1958; 166:1159-1172, 1294-1308. (A mortality analysis of 188,000 white males in nine U.S. States.)

Statistics are vital and extremely influential in almost every area of modern society—from politics to sports, housing to healthcare, ecology to entertainment—which helps to explain why being a statistician today is an extremely valued profession and 2013 bas been honored as the "International Year of Statistics."

2013: The International Year of . . . Statistics

by
Marie Davidian

It's a scene we statisticians know all too well. The passenger next to us asks, "So, what do you do?" "I'm a statistician," we reply, bracing for impact. "That was the worst course I ever took," our seatmate exclaims. "I had to take it for my major, but didn't understand any of it." It's a recurrent narrative we've experienced for decades.

But times have changed. From Google chief economist Hal Varian's well-circulated 2009 quote in *The New York Times*[1] that "the sexy job in the next 10 years will be statisticians" to the assertions of numerous blog posts[2] and articles[3], statistics is "hot."

And the dead-on predictions of the 2012 presidential election by not only statistician Nate Silver[4], but also Huffington Post's Mark Blumenthal and Simon Jackson[5], political scientist Drew Linzer[6], and others, have captured the attention of the public and media about the power of statistics like never before. Now, the reactions of our seatmates have transformed from groans into eager curiosity, due in large part to this watershed event.

What is statistics? To many, statistics is the class they took in college or figures on the sports pages. But statistics is so much more. Statistics is the science of learning from data and measuring, controlling, and communicating uncertainty. Statisticians do this by developing models to describe data. These models help us design methods to collect data, draw conclusions from data, and characterize the uncertainty in the findings.

For example, polls used by Silver and others to make their election predictions were designed by statisticians to learn about the true proportion of voters in a state favoring one candidate over another. In a poll, a sample of voters—selected to be similar in composition to the entire population of voters—yields a sample proportion that should reflect that for all voters. But it is likely a bit different, being based on only the sample, leading to uncertainty about the true proportion. Intuitively, the larger the sample, the less the uncertainty. Statistics provides tools for designing the sampling plan, quantifying the uncertainty, and determining how large a sample is needed to control the uncertainty below an acceptable threshold. Silver and his colleagues developed statistical models to combine the results of many polls to make their predictions and characterize the uncertainty in these.

Polls are a recognizable example of statistics in action. Most people don't know that statistics also benefits science and society in numerous ways. For instance, advances in medicine depend critically on statistics. For decades, statisticians have designed and analyzed clinical trials, the gold standard studies for comparing treatments. In the pharmaceutical industry, statisticians are involved in every step of drug development—from discovery through all phases of testing and clinical trials required for Food and Drug Administration approval.

And statisticians collaborate with medical and public health researchers on analyses of vast databases of patient information to uncover risk factors for disease, guide clinical practice, and generate new hypotheses.

Statisticians are working with medical and genomic scientists in the quest for personalized medicine, developing models of high-dimensional patient data to determine how to tailor treatment based on patient characteristics. The Environmental Protection Agency uses statistical modeling to uncover relationships between pollutants and illness and mortality and to establish regulatory standards.

Statistical models drive those purchase suggestions that pop up on Amazon and the ads that accompany Google searches. Businesses use statistics to develop marketing strategies based on Internet search and transaction data. Across the 14 federal statistical agencies, complex surveys—including the U.S. Census—are designed and conducted by statisticians. The results are used to determine congressional districts and allocate billions of dollars in resources for schools, health care, and transportation.

Scientific breakthroughs, such as the existence of the Higgs Boson[7] and discovery of the accelerating expansion of the universe[8]—awarded the 2011 Nobel Prize in Physics—used statistical models to establish that the findings were not simply artifacts of imprecise measurement.

Although of fundamental impact, these and myriad other contributions of statistics are widely unknown to the public, which is why more than 1,600 organizations in 112 countries are joining together to promote 2013 as the International Year of Statistics.

This global campaign features informational and educational resources and videos at the Statistics2013[9] website, and events are taking place worldwide throughout 2013.

The timing is ideal. The world is increasingly data driven and data dependent, and the avalanche of information being generated through online searches, electronic medical records, advances in genomic science, social networks, retail transactions, and remote sensing of the environment hold great potential for improving human health and business productivity and for guiding scientific discovery and public policy decision-making. Statisticians are essential to realizing the promise of this age of Big Data[10], ensuring that sound decisions are made and mitigating the threat of false discoveries.

It is this data deluge that has made statistics "sexy." A 2011 McKinsey Global Institute study[11] projects demand for 140,000 to 190,000 additional individuals with data analytic skills by 2018 in the U.S. alone, and 4 million worldwide. The opportunities for statisticians abound.

Modern applications of statistics have transformed the way it is taught. Meanwhile, interest in statistics is growing substantially among students. The number of high-school students taking the Advanced Placement Statistics exam has increased three-fold over the past decade, 7 percent in the last year alone. And the number of undergraduates majoring in statistics has risen dramatically. However, the number of graduate programs and PhDs, essential partners in cutting-edge scientific and business innovation, has grown only modestly. The demand for their skills in health sciences research alone will be significant, says a National Institutes of Health advisory group.[12]

So if your seatmate says, "I'm a statistician," ask about her work. Statistics will be more important than ever to scientists, policymakers, and you as we navigate the emerging data-rich era. As H. G. Wells purportedly said[13], "Statistical thinking will one day be as necessary for efficient citizenship as the ability to read and write."

Source

Davidian, Marie. (2013). "2013: The International Year of Statistics." Reprinted from *Huff Post: Science,* with the permission of the author. Originally posted 02/12/13 at http://www.huffingtonpost.com/marie-davidian/2013-the-international-ye_b2670704-.html.

Marie Davidian, Ph.D., is the 2013 President of the American Statistical Association and Professor of Statistics at North Carolina State University.

References

1. Lohr, Steve. (2009). For today's graduate, just one word: Statistics. *New York Times.* http://www.nytimes.com/2009/08/06/-technology/06stats.html?_r=1&

2. Hardy, Quentin (2012). What are the odds that stats would be this popular? *New York Times.* http://bits.blogs.nytimes.com-/2012/01/26/what-are-the-odds-that-stats-would-get-this-popular/

3. Teitell, Beth. (2012). The allure of the statistics field grows *The Boston Globe.* http://www.bostonglobe.com/lifestyle/style-/2012/11/21/

4. Silver, Nate. (2013). The White House is not a metronome."*New York Times.* (July 18). @ http://fivethirtyeight.-blogs.nytimes.com/

5. Jackman, Simon. (2012). Pollster predictions: 91.4% chance Obama wins, 303 or 302 EVs." http://www.huffingtonpost.-com/simon-jackman/pollster-predictions_b_2081013.html

6. Votamatic: (2012). Forecasts and polling analysis for the 2012 presidential election. http://votamatic.org/forecast-detail/

7. Getstats. (2012). Higgs Boson and the statistics of certainty. http://www.getstats.org.uk/2012/07/04/higgs-boson-and-the-statistics-of-certainty/

8. Perlmutter, S. et al. (1999). Measurements of Ω and Λ from 42 High-Redshift Supernovae. http://iopscience.iop.org/0004-637X/517/2/565/

9. The International Year of Statistics (2013). http://www.-statistics2013.org/

10. Lohr, Steve. (2012). The age of big data. *New York Times*: http://www.nytimes.com/2012/02/12/sunday-review/big-datas-impact-in-the-world.html?pagewanted=all&_r=1&

11. Manyika, J. et al. (2011). Big data: The next frontier for innovation, competition, and productivity. Insights and Publications: McKinsey Global Institute. http://www.mckinsey.com/insights/business_technology/big_data_the_next_frontier_for_innovation

12. National Institutes of Health. Advisory Committee to the Director. (2/12/11). http://acd.od.nih.gov/working-groups.htm

13. Quote: Wells/Wilks on statistical thinking. http://www.-causeweb.org/cwis/SPTFullRecord.php?ResourceId=1240

BLS 125th Anniversary

On the occasion of the 125th anniversary of the U. S. Bureau of Labor Statistics, the organization summarized a sample of the many statistical indicators, or "products," it produces that are essential for the functioning of the economy, and government in general, as described in this special "BLS Spotlight" report.

BLS 125th Anniversary

The Bureau of Labor Statistics (BLS) celebrated its 125th anniversary in 2009. In 1884, Congress voted to establish a Bureau of Labor, and on June 27 of that year President Chester A. Arthur signed the bill into law. A few years later the name of the organization was changed to Bureau of Labor Statistics. Eventually BLS became part of the Department of Labor when the Department was established in 1913.

In honor of this anniversary, BLS is shining the spotlight on a sample of its products.

Consumer Prices

The Consumer Price Index (CPI) is one of the oldest BLS products. The percent change in the index is often referred to as "the inflation rate." The CPI was initially begun to measure rates of inflation in ship-building centers during WWI, so that cost-of-living adjustments in wages could be calculated.

The CPI is now one of the most widely used measures of inflation in the U.S. Over 80 million individuals are affected by cost of living adjustments determined by the CPI, including Social Security beneficiaries. The CPI is also used to adjust the Federal income tax structure to prevent inflation-induced increases in taxes. In addition to overall inflation, the CPI program measures percent changes in prices for hundreds of individual consumer goods and services.

Producer Prices

BLS also measures changes in the prices received by producers of goods and services—Producer Price Indexes (PPI). Formerly called the Wholesale Price Index, the name was modified in 1978 to more accurately reflect the changing economy. . . . The three major producer price indexes are: finished goods (completed products ready for sale to retailers or the consumer); intermediate goods (commodities that have been processed but require further processing and nondurable, physically complete commodities purchased by business firms as inputs for their operations); and crude goods (farm products and raw materials). Producer prices are typically more volatile than consumer prices, particularly at the crude stage of processing.

Unemployment Rate

One of the most closely watched numbers from BLS is the national unemployment rate. It is released monthly, usually on the first Friday of the month. The rate is derived using responses from a sample of about 60,000 households; the data are collected by personal and telephone interviews. In addition to the overall unemployment rate, BLS calculates unemployment rates for many different groups of people—by age, race, gender, education, occupation, and so on.

Employment

Like the unemployment statistics, payroll employment figures from BLS garner much attention each month. These numbers come from a survey of about 150,000 businesses and government agencies, representing approximately 390,000 individual worksites. In addition to employment figures by industry, the payroll survey also yields hours and earnings statistics for many U.S. industries, as well as data for states and metropolitan areas. . . .

Employment Index

BLS began collecting information about worker pay shortly after its creation. For example, in 1888 the Bureau issued a study, "Working women in large cities," that reported on wages and other matters affecting working women. The most prominent BLS statistic pertaining to worker pay is now the Employment Cost Index (ECI). The ECI shows percent changes in employee compensation and its two components—wages and salaries, and benefits—with some of the numbers going back to the 1970s. The ECI is used for many purposes, such as adjustments to Medicare reimbursements for hospital, physician, and related services.

Productivity

Another prominent BLS statistic is productivity, which is measured as output per hour worked. Capital-intensive investment, improvements in technology, and better skilled workers are some of the major sources of labor productivity growth in the long term. Labor productivity has grown at an average annual rate of 2.2 percent in the nonfarm business sector during the past 60 years.

Injuries, Illnesses, and Fatalities Statistics

An early concern of the Bureau of Labor Statistics was measuring the rate at which workers are injured or become ill on the job. This is still an important area of study for BLS, and each year the Census of Fatal Occupational Injuries and the Survey of Occupational Injuries and Illnesses are conducted to provide the latest information on this topic. . . . The overall nonfatal injury and illness rate has declined in private industry in the U.S. since the early 1970s.

Job Openings and Labor Turnover Survey

One of the newest BLS programs is the Job Openings and Labor Turnover Survey (JOLTS). . . . This survey provides measures of job openings, hires, and separations . . . . The number of job openings is an important measure of the unmet demand for labor. When used in conjunction with the unemployment rate, it provides a more complete picture of the U.S. labor market. . . .

Geographic Data

In addition to national data, BLS publishes many statistics about states, metropolitan areas, and counties. For example, the unemployment rates for all of those types of geographic areas are available from the Local Area Unemployment Statistics program. . . .

Occupational Outlook Handbook

Although it's not a survey, the *Occupational Outlook Handbook* (*OOH*) is one of the best known products of the Bureau of Labor Statistics. First published shortly after World War II, the *OOH* has been used by generations of students to help them plan their careers. Each edition is packed with valuable information about hundreds of occupations, including their job tasks, working conditions, educational and training requirements, employment and wages, and prospects for the future.

Source

United States Department of Labor, Bureau of Labor Statistics, Spotlight on Statistics, June 2009 @ http://www.bls.gov/spotlight/2009/125_anniversary/home.htm

Highway systems are often called the "lifeblood" of an economy because they allow cargo and people to flow from place to place. When you see them, however, the last thing you are likely to think of is statistics—even though highway planning and operation depend fundamentally on statistics, as may be appreciated from this case study of the state of Arkansas.

The Necessity of Statistics in Highway Construction Management: The Case of Arkansas
by
Harold Rothbart

The Arkansas State Highway and Transportation Department is a large public service organization that strives to meet the transportation needs of the people within the entire state of Arkansas. To accomplish this objective, statistics are needed at virtually every stage of planning, production, and implementation. The department carries on a multitude of functions that broadly range from the determination of public needs to the conception and construction of public roads. Other functions carried out include planning, purchasing of highway rights of way, engineering design, public information, safety, construction, maintenance, weights and measure certification, and intergovernmental relations.

The Department is constantly seeking new methods, procedures and ideas to improve departmental operations, and previously instituted a maintenance management system, installed a new computer with increased capacity, purchased modern apparatus such as electronic distance measuring survey equipment, and completed a personnel classification and compensation study.

Another innovative method the Department used to improve services is the Construction Management Project, which sought to obtain more productivity, improve the scheduling of manpower needed for construction projects, and determine the amount of vehicles required at construction sites. For purposes of efficiency, economy, and production control, the Department is structured into three branches: Administrative and Realty, Planning, and Operations.

The statistics developed for Construction Management were concerned with the Construction Division within the Operations Branch. Its activities focused on producing and maintaining toll free roads at the minimal cost to all highway users, including the motoring public, the trucking industry, and agriculture for transport of farm-to-market products.

Typically, the Statistically Based Basic Scheduling System answered the following questions:

• How many people are required on the jobs assigned to a Resident Engineer Office?

• When will they be required and for how long?

• What quantity of hours will be needed to satisfy the workload?

• What is the relative need for personnel within the categories of inspection, surveying, documentation and supervision?

• Is there a surplus (or deficiency) in manpower? What can be done about it?

Specifically, the Statistically Based Scheduling System is a step-by-step process carried out by the Resident Engineer. It involves the periodic systematic analysis of each construction project assigned to the Resident Engineer to determine the resources (personnel) required to ensure adequate inspection of each project. The system results in a work schedule that outlines the personnel needs of the residency and assigns resources to meet those needs.

Variation by Season and Statistics

Construction activity in Arkansas is a function of season of year as well as the other factors. Due to inclement weather and seasonal limitations caused by frost, wet work conditions, or too much moisture in the soil for grading and excavation operations, the months of December, January, and February require less manpower than other months. This condition is true throughout all districts of the state, but is more severe in those located north of the central headquarters in Little Rock.

The seasonality of construction work must be taken into account when designing any statistically based management system. In order to obtain better work statistics for the variation of work by season, several years of data were analyzed. The data were obtained on man-hours worked in construction by Arkansas State Highway and Transportation Department personnel on a monthly basis.

Several generalizations were apparent from the statistical findings:

- The months of December, January, February, and March have reduced workloads compared to the rest of the year.

- The peak construction season consists of the six months, May, June, July, August, September and October.

- About 60 percent of the total manpower hour requirements occur during the peak six-month period.

- The average monthly man-hour requirements for the three years of data collected were: 51,471, 56,798, and 46,942.

Statistical Calculation of Total Man Hours for Each Project

To calculate the total man hours for each project, the Long-Range Forecasting System (LRFS) was used. It employs prediction equations to estimate the total man-hours required to inspect each construction project. These prediction equations were developed using a statistical technique called regression analysis. Regression analysis assumes that a definable relationship exists between the item that one desires to predict (the dependent variable, in this case man-hours) and other known project characteristics (the independent variables). Regression analysis then uses past (historic) information to define the relationship between the dependent and the independent variables. This defined relationship and the known information about the independent variables of future projects can then be used to predict the value of the dependent variable of those future projects.

In the specific case of the LRFS, the heads of the Construction Management Project wanted to predict the man-hours required to inspect any future construction project. It was believed that the man-hours required to inspect any future construction project were related to or were a function of certain characteristics of that project. This hypothesis was tested and verified.

The following characteristics that were believed related to the total man-hours required to inspect a construction project are:

Cost—the project contract cost.

Length—the length in miles of the roadway to be constructed.

ADT—Average daily traffic. This factor influences the number of lanes to be constructed and the thickness of the roadway to be constructed.These items influence the amount of construction work and therefore the amount of inspection man-hours.

Estimated Time—an estimate of the duration of time required to construct the project.

Type of Improvement—ten distinct groupings of types of improvements were identified, including Surfacing, Resurfacing, Grading and Drainage Structures, and Structural Approaches.

Terrain—the state was considered to be divided into three terrain categories: flats, rolling, and mountainous.

Urban/Rural—projects constructed in urban areas are slowed by control concerns and therefore require more inspection man-hours than would a similar rural project.

Functional Classification—roadways were divided into four categories: interstates, primary, secondary, and other.

Quarter—this variable refers to the quarter of the year in which a construction project was started. Quarters are labeled sequentially beginning in January/March.

Data were collected for each of these factors including the total inspection man-hours for 615 construction projects that were completed. The following is a sample calculation showing some of the variables involved:

Job Number	5891
Estimated Contract Cost	$300,000
Estimated Time	18 Months
ADT	4000
Estimated Start Date	January of Year 1
Estimated Completion Date	July of Year 2
Type of improvement	4
Functional Classification	1

The computer model would then make the appropriate substitutions and compute the total man-hours for a particular project.

As this essay has shown, statistics are necessary for planning, managing, and implementing the construction of highway and transportation systems. In the specific case discussed, statistical analyses improved the efficiency and effectiveness within the Arkansas State Highway and Transportation Department.

Source

This essay was originally written for the anthology.

Harold Rothbart holds a Bachelor of Civil Engineering degree, a Masters of Science degree in Engineering, and is a Licensed Professional Engineer in the states of New York and Michigan. He has had a varied career as a construction manager, management consultant, and assistant city administrator of Ann Arbor, Michigan.

Imaginative statisticians sometimes invent statistics with the intended purpose of improving societal institutions, as illustrated in this essay about a new statistic with the potential to protect against "gerrymandering" and advance equity during political elections.

Could a Statistic on Redistricting Have a Game-Changing Effect on American Politics?

by

Thomas R. Belin

People embrace statistics as a profession for a variety of reasons. For me, the field of statistics bridged broader interests in science and public policy, an ability in math, instincts to be precise in communication, lifelong enjoyment of sports, and a sense that we can make the world a better place by bearing witness. I see all of these themes as being intertwined with an idea that came to me as I was watching public-television coverage of Election Night 2004, pointing to the use of a statistic to advance the public interest by pushing back against corrosive polarization in American politics.

At the time, the 2004 presidential race between George W. Bush and John Kerry was hanging on returns from the close contest in Ohio. Meanwhile, a visual pattern routinely presented in televised election news coverage had come into sharper focus, with "red states" that had voted Republican dominating the southeast and western plains and "blue states" that had voted Democratic dominating the northeast and the west coast. Jim Lehrer, the dependable anchor and regular moderator of presidential debates, asked assembled panel members, "Why does the electoral map look the way it does with red states and blue states?"

The insight that triggered an avalanche of thoughts in me came from *New York Times* columnist David Brooks, who commented, "If you had to boil it down to one quantity, it would be housing density." Statistics is a discipline that thrives on learning from experience, reducing the complex to the simple, and drawing on information to communicate with one another. As my graduate-school advisor, Donald Rubin, used to say, "It's not an answer to say that the answer is in that stack of computer printouts over there—an answer is a story you can tell."

Part of what was so appealing about David Brooks's comment was that it not only explained patterns between states, with more urban states being blue and more rural states being red, but also patterns within states, with urban areas being blue and rural areas being red. Another appealing reality, known to me from my Ph.D. thesis project and related experience working at the U.S. Census Bureau, is that housing counts are routinely collected in the decennial census, which is the source of information used by states to draw up new congressional districts every decade consistent with the Supreme Court's "one-person/one-vote" rulings in the 1960s. And in that moment, I could see how a statistic reflecting population density could push back against the polarization of politics in modern America.

Going back to the early history of the United States, when Elbridge Gerry, whose redistricting plan for Massachusetts with a serpent-shaped district gave rise to the term "gerrymander" as a broader label for redistricting abuses, it has been widely recognized that the redistricting process provides opportunities for manipulating the political process. But polarization is seen as deriving more from politicians having "safe seats," where the lack of a credible threat of losing a future election results in their not having to compromise.

Political scientists have also long recognized that quite apart from the shapes of districts, it is possible to gain a partisan advantage by "packing" your opponents into a small number of districts where they enjoy a lopsided edge. As an example, in a state with seven districts where registered voters are evenly split between two parties but where representatives of one party control the redistricting process, it would be possible for partisan consultants making use of voter-registration data to devise five districts where the controlling party has a 60-40 edge, one district where the opposition party has a 70-30 edge, and one district with an 80-20 edge for the opposition party. By analogy, this would be like determining the champion of the National Basketball Association not based on the best-of-seven 5-on-5 games of basketball but, rather, allowing the team representing the conference that won the NBA All-Star Game to choose the number of players in each of seven games, subject only to the constraint that each game feature an average of five players to a side and each team would have an average of five players across all seven games. The outcome of the NBA finals, featuring five 6-on-4 games, one 3-on-7 game, and one 2-on-8 game would then essentially be determined by the outcome of the All-Star Game, which might be good for the controlling team but bad for everyone else, notably fans who value competition as a method of arriving at a winner.

But how could a statistic make a difference? Another analogy came to mind, namely with airline on-time arrival statistics. In the 1980s, a variety of disruptions in people's lives arose out of airline flight schedules with projected arrival times that were typically wildly optimistic. Calls for government regulation followed allegations that the airline schedules were intentionally misleading, as people tended to buy plane tickets through travel agents, and the agency computer systems would call up the shortest flights first. Instead of regulating the information directly, the government required airlines to publish information on the percentage of times that specific flights arrived on time, with parameters for determining what counted as an on-time arrival. The publication of this simple statistic had an immediate impact, leading to more accurately reported arrival times.

But based on a thought experiment that I pondered as the Ohio election results trickled in, devising a redistricting statistic seemed more complicated. I quickly discerned that elections would tend to be more competitive, at least in some districts, if there was less variation across districts in population density. But, as an example, if the goal were to balance population density across districts in New York State, one district could start on 10th Avenue in Manhattan and stretch all the way to Lake Erie, another district could start on 9th Avenue in Manhattan and stretch all the way to Lake Erie, and other similar "spaghetti-string" shaped districts balancing high-density and low-density areas of the state could be expected to give rise to a number of competitive elections. The problem, of course, is that such a framework would sacrifice another appealing feature of election districts, namely a sense of community or local-area representation. Any attempt to order redistricting plans from least favorable to most favorable would therefore have to balance multiple criteria, where a numerical summary favoring plans with less variation in population density across districts would need to be balanced by a numerical summary favoring plans where district shapes did not become too irregular.

What I eventually worked out with two graduate students in the UCLA Department of Biostatistics was what we called a "density-variation/compactness" or DVC score, where:

- *density variation* was measured as the average of the absolute values of the differences between district-specific population density (i.e., population divided by geographic area) and the overall average population density in the state, and

- *compactness* was measured as the area of a district divided by the area of a circle encompassing the district.

We published the details in an article in the journal *Statistics, Politics, and Policy* (2011), devising a scale where 0.0 corresponded to the redistricting plan in a state as of the year 2000; a score of 4.0 would be a high value (like a grade-point average), but it was possible for the score to be negative. We found that for California's 2002 redistricting, which governed over 850 legislative elections in the subsequent decade with only one seat changing party, the DVC score was -2.92; yet when we reconfigured the plan five different ways, we found three of the five had positive DVC scores, one with a score of 3.00. We also demonstrated that the DVC score we devised was favorably associated with a measure used by political scientists suggesting less partisan bias.

It is hard to discern how a redistricting plan is going to relate to the function (or dysfunction) in government just by looking at a redistricting map. Our instinct is that if DVC scores were routinely reported with proposed redistricting plans, there would be an outcry over plans with DVC scores deep in the negative range when positive scores, signaling more competitive districts, are so readily attainable.

Could this idea have a game-changing effect on American politics? I am encouraged by favorable feedback I have received from people with views ranging across the political spectrum, and there is time before the next decennial census for the idea to percolate, influence public opinion, and transform the politics of redistricting. Our government was established by leaders with tremendous vision, great faith in reason, great respect for individual citizens, and an awareness that political systems evolve. It would be a fitting tribute to them if a scientific insight, yielding a statistic as a one-number summary linked to certain desirable features of redistricting plans, would allow individual citizens to maintain control of a system of self-government that has vulnerabilities needing attention but that maintains its undeniable appeal.

Source

This essay was originally written for the anthology.

Thomas R. Belin, Ph.D., is Professor and Vice Chair, Department of Biostatistics, UCLA Jonathan and Karin Fielding School of Public Health.

Reference

Belin TR, Fischer HJ, Zigler CM. Using a density-variation/compactness measure to evaluate redistricting plans for partisan bias and electoral responsiveness. *Statistics, Politics, and Policy*, 2011; Volume 2: Article 3.

Statistics are essential for monitoring agricultural activity in the United States, including statistics on land usage, production, prices, income, wages, and other key variables, as this essay on "Ag" (agricultural) estimates from the National Agricultural Statistics Service (NASS) concisely explains.

The Importance of "Ag" Estimates

It would be hard to overestimate the importance of NASS's work or its contribution to U.S. agriculture. Producers, farm organizations, agribusinesses, lawmakers, and government agencies all rely heavily on the information produced by NASS.

Statistical information on acreage, production, stocks, prices, and income is essential for the smooth operation of Federal farm programs. It is also indispensable for planning and administering related Federal and State programs in such areas as consumer protection, conservation and environmental quality, trade, education, and recreation.

Moreover, the regular updating of information helps to ensure an orderly flow of goods and services among agriculture's producing, processing, and marketing sectors. Reliable, timely, and detailed crop and livestock statistics help to maintain a stable economic climate and minimize the uncertainties and risks associated with the production, marketing, and distribution of commodities.

Farmers and ranchers rely on NASS reports in making all sorts of production and marketing decisions. The reports help them decide on specific production plans, such as how much corn to plant, how many cattle to raise, and when to sell.

NASS estimates and forecasts are greatly relied upon by the transportation sector, warehouse and storage companies, banks and other lending institutions, commodity traders, and food processors.

Those in agribusiness who provide farmers with seeds, equipment, chemicals, and other goods and services study the reports when planning their marketing strategies.

Analysts transform the statistics into projections of coming trends, interpretations of the trends' economic implications, and evaluations of alternative courses of action for producers, agribusinesses, and policy makers. These analyses multiply the usefulness of NASS statistics.

Source

United States Department of Agriculture/National Agricultural Statistics Service. (2009). Education and Outreach: Importance of Ag Estimates. Retrieved 7/15/13: http://www.nass.usda.gov/Education_and_Outreach/Understanding_Statistics/Importance_of-_Ag_Estimates/index.asp

Statistics play a key role in improving public school educational programs, as explained in this essay about the types of data that are most useful to achieve school improvement, the procedures research personnel should follow to obtain and analyze data, the role of school executives in the process, and what will likely be gained through statistical analysis.

Data Inquiry and Analysis for Educational Reform
by
Howard H. Wade

Citizens and policymakers alike, as part of the new accountability, expect schools to justify the value and effectiveness of their programs. School boards, in turn, routinely ask to see the data administrators use to guide decision-making in schools. Boards are mindful, too, that allocation of state and federal funds often necessitates documented evidence that school programs lead to verifiable improvements in student achievement (Holcomb, 1999).

Apart from public relations and accountability issues, educators have come to recognize that they can no longer rely on "intuition, tradition or convenience" in making decisions about the best strategies to improve student learning (NCREL, 2000). For all these reasons, more schools across the country are settling on the idea that carefully collected and analyzed data represent the key to improvement in education.

This essay outlines the most useful types of data to drive the process of school improvement, the steps that must be taken to collect and analyze the data, the role of administrators in guiding the data-driven reform process, and the results that can be expected.

What Are the Educational Uses of Data?

Statistical data on school programs and student performance provide educators with their only real evidence of the success or failure of educational programs. Data "identify the link between teaching practices and student performance so that high achievement levels can be obtained" (Miller, 2000).

When systematically collected and analyzed, data provide an accurate way of identifying problem areas in school programs. Data reveal strengths and weaknesses in students' knowledge and skills, and they provide meaningful guidance on how teaching practices can and should be altered. When acknowledged and accepted by a faculty, data can lead to the formulation and implementation of corrective courses of action that can solve problems and meet a school's goals. Once improvement strategies are under way, educators can continue to analyze the data to monitor and refine their efforts.

From a wider perspective, data can "provide an honest portrayal of the district's and school's climate" (NCREL, 2000). Data can give a clear profile of a district and the schools within it: Who are our students, teachers, and families? What trends, attitudes, abilities, and values do they exhibit? What outside factors influence them?

What Types of Data are Most Useful?

The profile of a school or an educational program can include at least four types of data:

Student assessment data include measurements of student performance, such as standardized test results, grade point averages, and standard and other formal assessments, both state and federal.

Student demographic data describe such things as enrollment, attendance, grade level, ethnicity, gender, family background, and language proficiency. They may include facts on student mobility, modes of transportation, parent involvement, student behavior, and social problems.

Perceptions data document how a district or school is perceived not only by its students, teachers, and parents, but by the community at large. While this type of data is the most subjective—collected by questionnaires, surveys, interviews, and observations—it can often prove crucial in determining the direction a school may take in its programs and learning strategies.

School program data define programs, instructional strategies, and classroom practices. This type of data is most useful in monitoring, refining, structuring, and redirecting school programs. In collecting these data, "educators must systematically examine their practice and student achievement, making sure both are aligned with specifically defined, desired student outcomes" (Bernhardt, 2000).

How Can a School Begin Using Data?

The process of using data to improve school programs must be supported not just by one or two administrators, but by as many faculty members as possible. Data inquiry must grow out of a common recognition of the potential benefits statistical data can have in helping to achieve common goals.

Data analysis can be used at various levels within a school. Individual teachers can use it to improve teaching strategies in their classrooms. Groups of faculty can use data analysis to amend areas of concern within a department. Finally, the entire school can use data analysis to reform and improve the educational climate.

Although variations exist, depending on personnel and circumstances, educational reform through data analysis involves four steps.

First, the school staff identifies areas of concern for which improvement is desired. By gathering and analyzing existing assessment, demographic, perceptual, and program data, staff members create a profile of the school. This profile will reveal a school's strengths and weaknesses, and will allow interested faculty members, administrators, parents, and community members to ask appropriate questions regarding student achievement, teaching practices and programs, and school culture. Data collected during profiling can sometimes be used as baseline information against which subsequent improvement strategies can be measured.

Second, after the school profile is complete, interested teachers, administrators, and community participants meet together as a leadership team to prioritize the areas of concern that the profile has revealed. This is best achieved before the school year begins, and before faculty become immersed in the day-to-day challenges of teaching.

Third, once a problem is selected for correction, the leadership team collects and analyzes any additional data that might elucidate and suggest solutions to the problem. The team then outlines strategies for improvement and defines evaluation criteria against which the courses of corrective action will be measured.

Finally, after plans of action have been implemented and completed, the team again collects data to measure the success or failure of their efforts. The corrective strategies may be discontinued or adjusted and reapplied as needed.

Schools without experience implementing data analysis methods should start with simple problems, then advance to more complex issues only as benefits are witnessed and confidence gained. As well, it is important that the faculty commit to implementing the leadership team's recommendations, because without such commitment efforts for reform might end before benefits are realized.

For many schools, the appointment of a school-based facilitator or "data analyst" may be desirable. The analyst advises the leadership team in its collection and analysis of data, and in implementing corrective projects. In particular, the analyst helps "schools focus on their own particular problems, set locally appropriate goals, identify a course of action, and assess progress toward these goals" (Killion & Bellamy, 2000).

Selected from the faculty to encourage trust and confidence among staff for the process, the data analyst must be trained in data analysis, interpretation, and display processes. Because analysts are on-site, they can provide teachers and principals with timely responses and assistance (Killion & Bellamy, 2000).

How Can Administrators Overcome Barriers and Help Their Schools Use Data Effectively?

Despite the arguable benefits of data analysis, some educators remain ambivalent about its usefulness. No doubt a few educators cling to the antiquated notion that data collection is unprofessional. It certainly is true that, as Bernhardt notes, many educators "lack the training, equipment, and time to develop and carry out complex analyses." Last but not least, a fear often exists that data, if publicized, might somehow expose incriminating inadequacies and/or incompetence.

Principals and other administrators can best support data-driven reform processes by providing vision and leadership. Although they need not be proficient themselves in sophisticated data-gathering and data analysis techniques, they should be acquainted with the field, and should respect and value data analysis as an increasingly important tool in education.

Administrators do not necessarily need to direct the data collection and data analysis process, but they must support it. They must be able to assure the more skeptical participants that it is being undertaken not to blame or accuse, but to give both teachers and students the necessary tools to succeed.

Because administrators often possess a broader view of the issues that confront a school or district, they can help select the best, most easily measurable, results-oriented goals and initiatives. They can ask questions about student achievement and teaching practice, and can propose menus of initiatives based on the acquired data. Finally, they can guide staff development based on which initiatives have worked and which have not, according to analysis of the data.

What Results Can Be Expected from Data Inquiry and Analysis?

Clearly, data analysis is not a panacea that will solve every problem in a school. But when properly focused and implemented, data analysis is one tool that a school's staff can use to help raise educational achievement, thereby increasing confidence among faculty, students, and the community regarding the effectiveness of public education.

In learning to incorporate data analysis as a regular part of their professional activity, teachers become more reflective about their teaching practices, less reactive, less willing to accept easy answers, and more open-minded to solutions based on the data they gather. As a whole, the school assumes a more professional and civil culture of inquiry, in which "teachers share with each other important questions and ideas related to teaching and learning" (Feldman & Tung, 2001).

Through program improvements brought about by data analysis, a higher level of achievement can also be expected of students. In some cases, students have even begun emulating the practices of data inquiry they see their teachers modeling, conducting their own student surveys and analyses (Feldman & Tung, 2001).

Source

Wade, Howard H., (2001). Data inquiry and analysis for educational reform. ERIC Identifier: ED461911. ERIC Clearinghouse on Educational Management, Eugene, OR. http://files.eric.ed.gov/fulltext/ED461911.pdf

References

Bernhardt, Victoria L. "New Routes Open When One Type of Data Crosses Another." Journal of Staff Development 21, 1 (Winter 2000). http://www.nsdc.org/library/jsd/bernhardt211.html

Calhoun, Emily F. How To Use Action Research in the Self-Renewing School. Alexandria, Virginia: Association for Supervision and Curriculum Development, 1994.

Chrispeels, Janet H., and others. "School Leadership Teams: A Process Model of Team Development." School Effectiveness and School Improvement 11, 1 (March 2000): 20-56. EJ 611 222.

Du, Yi, and Larry Fuglesten. "Beyond Test Scores: Edina Public Schools' Use of Surveys To Collect School Profile and Accountability Data." ERS Spectrum (Summer 2001): 20-25.

Feldman, Jay, and Rosann Tung. "Using Data-Based Inquiry and Decision Making to Improve Instruction." ERS Spectrum (Summer 2001): 10-19.

Holcomb, Edie L. Getting Excited About Data: How To Combine People, Passion, and Proof. Thousand Oaks, California: Corwin Press, 1999. 162 pages. ED 433 373.

Killion, Joellen, and G. Thomas Bellamy. "On the Job: Data Analysts Focus School Improvement Efforts." Journal of Staff Development 21, 1 (Winter 2000). http://www.nsdc.org/library/jsd/killion211.html

McLean, James E. Improving Education Through Action Research: A Guide for Administrators and Teachers. Thousand Oaks, California: Corwin Press, 1995. 87 pages. ED 380 884.

Miller, A. Christine. "School Reform in Action." Paper presented to the American Educational Research Association Conference, New Orleans, April 28, 2000.

North Central Regional Educational Laboratory. Using Data To Bring About Positive Results in School Improvement Efforts. Oakbrook, Illinois: Author, December 2000.

Sparks, Dennis. "Results Are the Reason." Journal of Staff Development 21,1 (Winter 2000).

Many statisticians derive great pleasure from their occupation, a fact that is clearly conveyed in this interview conducted by Kristi Birch with a leading biostatistician, who discusses her work assessing the health consequences of air pollution, as well as what it's like being a woman in academia and balancing work and family life.

Interview with Francesca Dominici

Francesca Dominici, Ph.D., analyzes mega-databases to assess the impact of air pollution levels around the U.S. on the incidence of heart disease, stroke, and death. Here, she talks about public health research, public health data, and the challenge of explaining public health data to Congress.

Were you interested in math as a kid?

Yes, I loved math! Doing math was equivalent to having fun. Solving a hard math problem gave me a lot of satisfaction. And I always loved probability. It was challenging but at the same time fun and intellectually stimulating.

When did you get interested in biostatistics in particular?

At the beginning of my PhD in statistics. I really wanted to use statistics to solve hard problems in science. My advisor proposed working on a project that compared the efficacy of several treatments for migraine headaches from heterogeneous clinical trials. The clinical trials were heterogeneous in the sense that they used different sampling designs, not always including a placebo and often comparing the efficacy of different sets of treatments. I enjoyed developing complex statistical models that properly integrated the information across the heterogeneous trials and provided a comprehensive ranking of the treatments.

What other kinds of researchers do you work with?

Epidemiologists, doctors, basic science scientists, and computer scientists.

What's fun about your job?

Learning about important scientific problems, identifying what evidence is needed, and figuring out the best way to analyze massive amounts of information. In addition, a biostatistician is trained to quantify the degree of uncertainty in his/her estimated value.

What's a good day at work for you? What's a bad day?

A good day is filled with meetings with students and colleagues where we make progress on different projects. A bad day is filled with bureaucratic work. Fortunately, because of our excellent staff, there are very few bad days.

You're interested in using your data to effect policy change, and you've testified before the EPA and other groups. Data can seem pretty dry to some people. How do you present your findings in a way that engages your audience?

This is the most challenging and satisfying part of my job. I try to communicate our findings as clearly and transparently as I can. Our data are better understood and remembered if they are translated into the number of human lives that could be saved by lower levels of air pollution. So I talk about that, rather than just presenting data about what constitutes harmful levels of air pollution. I also usually use PowerPoint slides to present the data visually.

You have presented your data on air pollution to Congress. If the government does decide to take action, what goals would need to be achieved to reduce the harmful effects of air pollution?

Setting specific goals to reduce air pollution is not my job; my job is to use the best available data and develop methods to provide evidence as to whether or not current levels of air pollution are harmful. But my work is important to reducing pollution, because the more compelling the evidence about air pollution, the more likely that it will result in public health action.

You've found that ground-level ozone pollution increases the number of deaths and that short-term exposure to particulate matter in the air increases the number of hospitalizations for cardiovascular and respiratory disease. Did either of these results (or the magnitude of them) surprise you?

Yes, these results did surprise me. Although I suspected that this kind of air pollution was harmful, I did not expect that we would be able to detect such small effects in the presence of so many factors, such as weather and seasonality, that could mask the association.

What is the hardest part of your job?

Always working under a deadline and switching to a different project every hour.

Where do you get all your data and how do you get it? Are the data expensive?

We use national government data, like the billing claims from Medicare data, deaths from the National Center of Health Statistics, Air pollution from the EPA, and weather from NOAA. The air pollution and weather data are free. The Medicare data can be very expensive. . . .

What kind of computer software do you use in your work?

We use SAS and R. These are the common statistical software packages for biostatisticians. R is free software and is used more often for biostatisticians that develop new methods. SAS is widely used in the industry for existing statistical approaches.

If you weren't a biostatistician, what else would you like to be?

Probably a physician, doing a combination of clinical and research work.

Is there anything in your career that you wish you'd done differently?

No.

Who has influenced you most in your work and how?

My husband, Dr. Giovanni Parmigiani, who is also a biostatistician, and my two mentors, Jonathan Samet and Scott Zeger. Dr. Samet helped me appreciate the importance of working in a controversial environment and taught me how to use statistics to make a change. Dr. Zeger helped me better understand how to analyze data to address a particular scientific question and is teaching me to be a leader. My husband has always been there to help me overcome challenges.

If you weren't working at a school of public health, where else could you use your skills as a biostatistician?

In all academic environments and also research centers such as the Center for Disease Control and Prevention (CDC), the World

Health Organization (WHO), the Environmental Protection Agency (EPA), and the National Institutes of Health (NIH).

Right now we have too much data but not enough biostatisticians. When a large amount of information is not processed adequately, it leads to more confusion, not to more knowledge.

How do you balance work and home?

It is very hard. I have an 11-month-old daughter. There are still serious gender-based obstacles in academia. In an environment where the ideal worker works 24/7, reaching a balance is hard. Most of the men academic leaders have a wife at home. But how about the women?

What advice do you have for kids wanting to go into biostatistics? What should they be doing now?

Get excellent training in mathematics and probability, as well as a good exposure to scientific problems. They can always contact our biostatistics program and come for a visit!

Source

This interview is reproduced with the permission of Francesca Dominici. It originally appeared in *Johns Hopkins Public Health* (2003) under the title "Generation Next" (http://magazine.jhsph.edu/2003/fall/generation_nxt/dominici.html), and subsequently on cogito.org, as "Interview with Francesca Dominici," https://cogito.cty.jhu.edu/11136/interview-with-francescadominici-biostatistician-2/, a site of John's Hopkins University.

Francesca Dominici is a Biostatistician and Associate Dean for Information Technology, Harvard University, School of Public Health.

Research scientists often experience great pleasure when they test hypotheses and obtain statistically significant results, especially when the findings can positively affect many individuals and stimulate further investigation, as this essay on obesity research illustrates.

The Joy of Research Discovery with Statistics

by

Allan Geliebter

As a psychologist, I have been conducting research on obesity, a topic which has been receiving considerable attention recently with the worldwide obesity epidemic. My research in obesity relies greatly on statistical analysis, and I have been fortunate to have had solid training in statistics. Although scientific discoveries happen only occasionally in a scientific career, when they do, they bring much joy. I have had the privilege of making some novel discoveries and will share one with you.

We had previously shown in my lab that obese individuals have a much larger stomach capacity than lean individuals (Geliebter, 1988). This finding was made possible by a technique we had developed to measure stomach capacity. The technique involves filling a balloon inserted into the stomach with water through a tube until the person rates abdominal discomfort as "10" on a scale from "1" to "10." The volume reached at the rating of 10 is considered the person's stomach capacity.

I then wondered if stomach capacity could change, and specifically whether capacity would be reduced by eating smaller portions of food while dieting to lose weight. The popular notion that dieting would make the stomach "shrink" had not been tested scientifically. We measured stomach capacity in two groups of obese subjects prior to and after a 4-week period: 1) 14 who went on a low-calorie diet and 2) 9 weight-matched controls who did not go on a diet (Geliebter et al., 1996). When the data on all the subjects had been collected, I found a statistically significant reduction in stomach capacity, based on the subjective measurement of discomfort, of 27 percent in the diet group. I remember my excitement about the finding and staying in the lab late into the night to recheck the statistical analyses. I went home before I could analyze the results from the control group.

The importance of analyzing the data from the control group was to ensure that the findings were not just a result of repeated exposure by the diet group to the testing procedure the second time around, referred to as a practice effect. The next day, I confirmed that the stomach capacity of the control group did not change significantly.

I still remained somewhat concerned that our results showing a reduction of stomach capacity were based up to this point on a subjective sensation of discomfort. Perhaps people who are dieting become more sensitive to the discomfort of a balloon distending the stomach, and therefore reach a rating of 10 at a lower volume, which incorrectly indicates reduced stomach capacity. I therefore also examined another more objective index of stomach capacity, intentionally added to the study, based on pressure readings recorded from within the stomach while the balloon was being filled. This other measure could also reflect stomach capacity, based on the principle that a stomach with a larger capacity requires a greater balloon volume to produce a specific rise in pressure than a stomach of smaller capacity. When I examined the pressure data from the stomach, I obtained similar results, namely that the volume needed to produce a given rise in pressure was significantly reduced after 4 weeks only in the dieting group and not in the control group, confirming the initial findings.

In the next several days, I also was able to rule out the possibility that the weight change from the diet (a potential confounding factor) was contributing more to the findings than the reduced food intake during the diet. I showed that the relationship between the weight loss and the change in stomach capacity was not statistically significant.

I was also pleased that after the study was published it continued to stimulate more scientific research, including a recent brain imaging study in collaboration with colleagues at Brookhaven National Labs (Wang et al., 2011). We showed that distending the stomach of people while inside an MRI scanner excited two regions of the brain, the insula, which receives sensations from internal organs, and the amygdala, which is involved in emotion and reward. This study also relied highly on statistics, mainly to help isolate the specific brain regions that were responding to stomach distension from tens of thousands of potential brain areas.

Although the finding of reduced stomach capacity following a diet was by no means an epic discovery, it received more media attention than any other research discovery I have made. The likely reason is that, unlike much of my research which is quite technical, this finding was relatively easy to understand by lay people. They could also relate it to their own experience when following a diet of feeling full sooner after eating the same amount of food as before the diet. Of course, this added attention to the study findings was gratifying and enhanced my joy of scientific discovery.

Source

This essay was originally written for the anthology.

Allan Geliebter, Ph.D., is Senior Research Scientist, Department of Psychiatry, Columbia University Medical Center, and Professor, Department of Psychology, Touro College, New York, NY.

References

Geliebter A. Gastric distension and gastric capacity in relation to food intake. *Physiology and Behavior* 1988, *44*:665-668.

Geliebter A, Schachter S, Lohmann C, Feldman H, Hashim SA. Reduced stomach capacity in obese subjects after dieting. *Am J Clin Nutr* 1996, *63*:170-173.

Wang GJ, Tomasi D, Backus W, Wang R, Telang F, Geliebter A, Korner J, Bauman A, Fowler JS, Thanos PK, Volkow ND. Gastric distention activates satiety circuitry in the human brain. *Neuroimage*, 2008, Feb 15, *39*(4):1824-31.

Statistics have played an important role in helping to reduce the incidence of cigarette smoking by onscreen characters in youth-oriented movies, monitoring motion picture studios' depictions of onscreen smoking, and promoting legislation to eliminate onscreen smoking, as described in this article from the U.S. Centers for Disease Control.

Smoking in Movies
by
Tim McAfee and Michael Tynan

Youth who are heavily exposed to onscreen smoking are approximately 2 to 3 times as likely to begin smoking as youth who are lightly exposed (1), and the Surgeon General concluded that there is a causal relationship between depictions of smoking in the movies and smoking initiation among young people (2). Among the 3 major motion picture companies with policies aimed at reducing tobacco-use incidents in their movies, the number of onscreen incidents per youth-rated movie (rated G, PG, or PG-13 by the Motion Picture Association of America) decreased 95.8% from 2005 through 2010 (3). These results appeared to indicate that movie companies were making progress at reducing smoking depictions in youth-oriented movies and that a company-by-company approach of adopting voluntary policies could be effective in nearly eliminating youth exposure to tobacco imagery in movies. However, new data from 2011 published by Glantz and colleagues (4) in *Preventing Chronic Disease* raise serious concerns about this individual company approach.

Glantz and colleagues found that in 2011, depictions of tobacco use per youth-rated movie rebounded; estimated instances of tobacco use in 2011 were more than one-third higher than in 2010

(4). Furthermore, the authors found that the largest increase in tobacco-use incidents in youth-rated movies was among the 3 movie companies that had produced the dramatic decline from 1995 through 2010 and had policies designed to discourage depictions of smoking in their movies. As a result of this sharp rebound, the difference in tobacco-use incidents per youth-rated movie between companies with policies and companies without policies diminished in 2011 (4). This difference suggests that individual company policies may not be sufficient to sustain a reduction in youth exposure to tobacco-use and other pro-tobacco imagery in movies and that more formal, industry-wide policies are needed.

The World Health Organization and other public health groups have recommended formal policies aimed at eliminating smoking in the movies (5, 6). These policies include awarding an R rating to any movie with smoking or other tobacco-use imagery (with exceptions for portrayal of actual historical figures who smoked or the portrayal of negative effects of tobacco use), certifying that no payments have been received by studios for depicting tobacco use in movies, and ending the onscreen depiction of real tobacco brands.

Reducing smoking and tobacco use in youth-oriented movies is not a niche issue. The Surgeon General has concluded that there is a causal relationship between depictions of smoking in movies and smoking initiation among young people, and the US Department of Health and Human Services has set a goal of reducing youth exposure to onscreen smoking (7). Furthermore, among the nationwide goals set by Healthy People 2020, one of the objectives is the reduction of onscreen tobacco use imagery in youth-oriented movies and on television (8). These goals and objectives were set because the population-attributable risk associated with onscreen tobacco imagery is significant (9, 10).

To assess progress toward the Healthy People 2020 objective, the Office on Smoking and Health at the Centers for Disease Control and Prevention (CDC) will now track and report annually on tobacco use imagery in youth-oriented movies as a core surveillance indicator by using the methods described in previous CDC publications (3, 4). These data will be added to regular CDC reports to the public on smoking prevalence among youth and adults, total and per-capita cigarette consumption, and progress on tobacco control policies.

One of the major conclusions in the Surgeon General's 2012 report on preventing tobacco use (2) was that after years of steady progress, declines in tobacco use by youth and young adults have slowed for cigarette smoking and have stalled for smokeless tobacco use (2). Each day in the United States, approximately 3,800 young people younger than 18 years smoke their first cigarette, and approximately 1,000 youth younger than 18 years become daily cigarette smokers (2). More than one-third of these smokers will eventually suffer and die from smoking-related illness. We all have a responsibility to prevent youth from becoming tobacco users, and the movie industry has a responsibility to protect our youth from exposure to tobacco use and other pro-tobacco imagery in movies that are produced and rated as appropriate for children and adolescents. Eliminating tobacco imagery in movies is an important step that should be easy to take.

Source:

McAfee, T., MD, MPH & Tynan, M. Smoking in Movies: A New Centers for Disease Control and Prevention Core Surveillance Indicator. Prev Chronic Dis 2012;9:120261.DOI: http://dx.doi.org-/10.5888pcd9.120261.

References

1. National Cancer Institute Tobacco control monograph 19: the role of the media in promoting and reducing tobacco use. Bethesda (MD): US Department of Health and Human Services, National Institutes of Health, National Cancer Institute; 2008.

2. A report of the Surgeon General: preventing tobacco use among youth and young adults: a report of the Surgeon General, 2012. Atlanta (GA): US Department of Health and Human Services, Centers for Disease Control and Prevention, National Center for Chronic Disease Prevention and Health Promotion, Office on Smoking and Health; 2012.

3. Centers for Disease Control and Prevention (CDC) Smoking in top-grossing movies United States, 2010. MMWR Morb Mortal Wkly Rep 2011;60(27): 910–3.

4. Glantz SA, Iaccopucci A, Titus K, Polansky JR. Smoking in top-grossing US movies, 2011.Prev Chronic Dis 2012;9:E150. doi: 10.5888/pcd9.120170.

5. Sargent JD, Tanski SE, Gibson J. Exposure to movie smoking among US adolescents aged 10 to 14 years: a population estimate. Pediatrics 2007;119(5):e1167–76. doi: 10.1542/peds.-2006-2897

6. World Health Organization Smoke-free movies: from evidence to action. Geneva (CH): World Health Organization; 2009. http://www.who.int/tobacco/smoke_free_movies/enAccessed 10/18/2012.

7. Ending the tobacco epidemic: a tobacco control strategic action plan for the US Department of Health and Human Services. Washington (DC): US Department of Health and Human Services; 2010.

8. Tobacco use. In: Healthy people 2020. Washington (DC): US Department of Health and Human Services; 2010.

9. Sargent JD, Tanski S, Stoolmiller M. Influence of motion picture rating on adolescent response to movie smoking. Pediatrics 2012;130(2):228–36. doi: 10.1542/peds.2011-1787.

10. Millett C, Glantz SA. Assigning an "18" rating to movies with tobacco imagery is essential to reduce youth smoking. Thorax 2010;65(5):377–8Erratum in: Thorax 2010;65(9):844. doi: 10.11-36/thx.2009.133108.

Statistics have revolutionized sports in many ways, from how ballgames and other competitions are announced and reported on mass media to how managers and owners recruit and assemble teams using the statistical science of "sabermetrics," as this essay explains about the most prominent example in sports, the case of the Oakland Athletics.

Statistics, *Moneyball* and "Sabermetrics"

In 2003 Michael Lewis published *Moneyball: The Art of Winning an Unfair Game*, which became a movie starring Brad Pitt in 2011. The main premise of *Moneyball* is that the wisdom of all key baseball personnel and insiders over the past century—including players, managers, coaches, scouts, and front office staff—is subjective and, therefore, often erroneous. Statistics on key indicators such as stolen bases, RBIs (runs batted in), and batting averages that scouts and managerial personnel usually use to assess players are vestiges of the nineteenth century based on the only statistics available at the time and, therefore, reflect an outdated understanding of player performance.

All of this changed radically, however, as Lewis describes in *Moneyball,* when the general manager of the Oakland A's, Billy Beane, and others in the A's front office identified different measures of player performance using *sabermetics*—the specialized analysis of baseball through statistics and other objective evidence that measure players' in-game activities. The term "sabermetrics" is derived from the acronym SABR, which stands for the Society for American Baseball Research. It was coined by Bill James, a legendary baseball statistician, writer, and historian, whose seminal *Baseball Abstract*—an annual reference work published from the late 1970s through the late 1980s—influenced many of the young, up-and-coming baseball minds now joining the ranks of baseball management, and gave rise to such websites based on sabermetrics as *Baseball Prospectus*, which Nate Silver describes as "a stat geek's wet dream" (2012, p. 78).

Based on these new statistical measures derived through sabermetrics, manager Beane and the A's front office executives assembled a team that was able to compete successfully against the New York Yankees and other much richer Major League Baseball organizations.

More specifically, through rigorous statistical analysis the A's demonstrated, for example, that on-base percentage and slugging percentage are better indicators of offensive success; and the A's, therefore, became convinced that players with these qualities were cheaper to obtain on the open market than players with more historically valued qualities such as speed and contact. These observations often flew in the face of conventional baseball wisdom and the beliefs of many baseball scouts and executives.

By re-evaluating the strategies that produce wins on the field, the 2002 Athletics, with approximately $41 million in salary, were competitive with larger market teams such as the New York Yankees, who spent over $125 million in payroll that same season. Because of the smaller revenues of the A's, Oakland was forced to find players undervalued by the market, and their system for finding value in undervalued players has proven itself thus far. This approach brought the A's to the playoffs in 2002 and 2003.

In time, other teams began to use sabermetrics to mirror Beane's strategies for evaluating offensive talent, which diminished the Athletics' advantage and forced them to begin looking for other undervalued baseball skills such as defensive capabilities.

After the 2010 season, when the New York Mets hired Sandy Alderson—Beane's predecessor and mentor with the A's—as their general manager, and hired Beane's former associates (Paul

DePodesta and J. P. Ricciardi) to the front office, the team became known as the "Moneyball Mets."

Moneyball has made such an impact in professional baseball that the term itself has entered the lexicon of baseball. Teams which appear to value the concepts of sabermetrics are often said to be playing "Moneyball." Baseball traditionalists, in particular some scouts and media members, decry the sabermetric revolution and have disparaged "Moneyball" for emphasizing concepts of sabermetrics over more traditional methods of player evaluation. Nevertheless, the impact of "Moneyball" upon major league front offices is undeniable. In its wake, teams such as the New York Mets, New York Yankees, San Diego Padres, St. Louis Cardinals, Boston Red Sox, Washington Nationals, Arizona Diamondbacks, Cleveland Indians, and the Toronto Blue Jays have hired full-time sabermetric analysts.

References

Lewis, Michael. (2003). *Moneyball: The art of winning an unfair game.* New York: W. W. Norton & Company.

Silver, Nate. (2012). *The Signal and the Noise: Why so many predictions fail—but some don't.* New York: The Penguin Press.

Wikipedia. (2013). *Moneyball: The art of winning an unfair game.* http://en.wikipedia.org/wiki/Moneyball.

Data-driven, or statistical, decision making is particularly important in urban schools whose populations are disproportionately poor, minority, and in need of special services. This essay discusses the types of data that schools should collect and the ways to use the information effectively in decision making to enhance equity.

Data-Driven Equity in Urban Schools
by
Wendy Schwartz

Data Use to Enhance Equity

Collecting and analyzing meaningful data about the characteristics and academic performance of students, and about school organization and management, helps under-resourced, underperforming, and highly diverse schools "identify achievement gaps, address equity issues, determine the effectiveness of specific programs and courses of study, and target instructional improvement" (Lachat, 2002, p. 3). The process helps provide individualized instruction and services to meet the needs of a single student or a specific group of students. It also enables a school continually to monitor the effects of its efforts to determine whether changes are needed. Thus, a school can help ensure the success of the students who most need special supports and who otherwise might not have been noticed.

Data Types

Conducting an overall evaluation to see how the school is doing is a good way to begin data collection. This process that helps school administrators, counselors, and teachers explore, discover, and address the individual needs of the school and students (Slowinski, 2002; Wade, 2001). Four basic categories of data are important to collect (Bernhardt, 2001; Lachat, 2002; Slowinski, 2002):

Student Learning Data: These data comprise information on individual students. They include, for example, prior schools attended, courses taken, and achievement; current classroom assessments, grades, standardized test scores, and comparisons among them; diagnostic assessments; programs enrolled in and courses taken (i.e., special education, bilingual education, school-to-career); and participation in projects and extracurricular activities. The resulting longitudinal data provide information about a student over time and identify factors that may have influenced performance (i.e., a school transfer, a change in program). Aggregating student learning data provides a profile of the school, such as the total student group at the top level of achievement or those receiving tutoring.

Student Demographics: These data comprise personal factors about each student: gender, ethnicity, socioeconomic level, language proficiency, and other equity factors. They also comprise characteristics related to a student's schooling, such as attendance and discipline record, and school mobility.

Perceptions Data: These data provide information about perceptions of school personnel, parents, and the community; about the school at large; about the classes; and about specific programs and strategies.

School Process Data: These data provide information about school management, administration, organization, and operations, including the programs the school offers, classroom strategies, and instructional practices.

Data Analysis

Disaggregating the data by breaking them down into smaller elements allows a school to determine more accurately the effects of its programs and strategies on segments of its student body. To conduct an initial equity analysis, for example, the students taking advanced courses can be disaggregated by ethnicity, gender, and socioeconomic status to ascertain the percentage, by group, enrolled in those courses and to consider whether placement decisions are biased or whether the achievement of some groups is routinely lower than that of others (Lachat, 2002).

To determine whether specific programs are not working for certain students, underachieving students, students with excessive absenteeism, and dropouts, data can be disaggregated by courses taken, or even by the teachers who taught them, to consider relevant school influences. The performance of students with similar personal characteristics can be disaggregated to determine which courses, instructional strategies, etc., are most effective with them. Differences between student grades and scores on standardized tests can be reviewed to determine whether there are lags in course content or poor preparation for some types of tests (Lachat, 2002). Finally, item analysis—the process of examining the group of students that missed a particular item on one test or similar items on several assessments—can determine what, if any, factors they have in common, such as the same teacher or limited opportunity to learn the material (Slowinski, 2002).

Data-Driven Decision Making

Disaggregating data uncovers interesting realities that can form essential questions about the reasons for student outcomes and the ways to reverse poor results. These questions should drive the investigation phase that involves, for example, researching best practices and proven programs for students like those of the school. After identifying the programs or program components, and the practices likely to meet the needs of its students, the school is ready to develop an implementation plan. A pilot with a small group can determine unintended consequences as well as the impact of the intervention (Slowinski, 2002).

Choosing a Technology Tool to Support Data-Driven Decision Making

A technology tool is required to manage data flow. Several factors need to be considered in selecting one (Slowinski, 2002):

Functionality. At a basic level, a tool should be able to import data from a variety of electronic sources into a relational database that can disaggregate the data. The tool should also be able to generate a report easily through graphs and diagrams of the data and their relationships.

Data Storage Capacity. The technology tool must accommodate the number of students in the district or school and the amount of data collected over the time period determined.

Training. The school needs to ascertain how easily the tool can be learned, whether an individual at the local level will need to commit a significant amount of time to learning and using it, the availability of such an individual, and the cost of training and technical assistance.

Format. Decisions, to some extent based on relative costs, should be made about whether to purchase a tool or outsource its use to a vendor, and where to house the data (at the district or school or with the vendor). Schools also need to consider the format of the tool (i.e., web-based), mode of access (i.e., via the Internet), ease of access of school personnel, and security issues.

Source

Schwartz, Wendy. (2002). Data-driven equity in urban schools. Adapted from ERIC Digest, ERIC Identifier: ED467688. ERIC Clearinghouse on Urban Education New York NY. http://files.eric.ed.gov/fulltext/ED467688.pdf

References

Bernhardt, V.L. (2001, Winter). Intersections. Journal of Staff Development, 21(1), 33-36. (EJ 600 392) Available at: http://www.nsdc.org/library/jsd/bernhardt211.html

Lachat, M.A. (2002). Data-driven high school reform: The breaking ranks model. Hampton, NH: Center for Resource Management.

Slowinksi, J. (2002). Data-driven equity: Eliminating the achievement gap and improving learning for all students. Unpublished manuscript, Vinalhaven Schools, Vinalhaven, ME.

Wade, H. H. (2001, December). Data inquiry and analysis for educational reform. ERIC Digest 153. Eugene, OR: ERIC Clearinghouse on Educational Management. Available at: http://eric.uoregon.edu/publications/digests/digest153.html

Industries vital to the United States economy are tracked using a battery of statistical indicators on many variables, including consumer spending, employment, and productivity, as this "Spotlight" article from the Bureau of Labor Statistics details, with projections through the year 2020.

Statistics and the Fashion Industry

Throughout history, fashion has greatly influenced the "fabric" of societies all over the world. What people wear often characterizes who they are and what they do for a living. As Mark Twain once wrote, "Clothes make the man. Naked people have little or no influence on society."

The fashion industry is a global industry, where fashion designers, manufacturers, merchandisers, and retailers from all over the world collaborate to design, manufacture, and sell clothing, shoes, and accessories. The industry is characterized by short product life cycles, erratic consumer demand, an abundance of product variety, and complex supply chains.

In this Spotlight, we take a look at the fashion industry's supply chain—including import and producer prices, employment in the apparel manufacturing and fashion-related wholesale and retail trade industries, labor productivity in the manufacturing sector and in selected textile and apparel industries, and consumer prices and expenditures on apparel-related items.

How Much Do Consumers Spend on Apparel?

In 2010, households spent, on average, $1,700 (in nominal terms) on apparel, footwear, and related products and services—3.5 percent of average annual expenditures. Since 1985, as a percentage of total apparel expenditures, households spent more, on average, on apparel designed for women aged 16 and over than any other apparel product or service.

Employment in Apparel Manufacturing

Employment in the apparel manufacturing industry has declined by more than 80 percent (from about 900,000 to 150,000 jobs) over the past two decades. The decline has been proportional throughout the apparel manufacturing component industries.

Where in the United States is Apparel Made?

The apparel manufacturing industry includes a diverse range of establishments manufacturing full lines of ready-to-wear and custom apparel; apparel contractors, performing cutting or sewing operations on materials owned by others; and tailors, manufacturing custom garments for individual clients. Knitting, when done alone, is classified in the textile mills subsector, but when knitting is combined with the production of complete garments, the activity is classified in the apparel manufacturing industry.

In 2010, there were 7,855 private business establishments in the apparel manufacturing industry, employing 157,587 workers—compared with 15,478 establishments and 426,027 workers in 2001. In 2010, only two U.S. counties had more than 500 business establishments—Los Angeles county, California (2,509) and New York county, New York (803).

Establishment Size

The average size of establishments (the number of employees at a typical workplace such as a factory or store) has declined in most apparel manufacturing industries in recent years, while it generally remained little changed in fashion-related retail trade industries.

In apparel manufacturing, the average number of employees per establishment declined from 28 to 20 over the 2001–2010 period, though it stayed about the same in women's and girls' cut and sew apparel manufacturing.

The average number of employees per establishment in clothing stores stayed near 13 during the 2001–2010 period, though it decreased from 25 to 21 in family clothing stores.

Wholesale and Retail Trade Employment

From 1990 to 2011, within the wholesale trade industry, employment in industries such as jewelry and women's and children's clothing experienced little or no change. However, over that period, employment in the men's and boy's clothing industry decreased 17.5 percent—from 32,000 jobs in 1990 to 26,400 jobs in 2011.

Within the retail trade industry, employment in men's and women's clothing stores, shoe stores, and jewelry, luggage, and leather goods stores decreased from 1990 to 2011. In contrast, industries such as children's and infant's clothing (118.6 percent), cosmetic and beauty supply stores (82.3 percent), family clothing (63.2 percent), and clothing accessories stores (57.0 percent) all experienced an increase in employment from 1990 to 2011. From 1990 to 2007, employment in family clothing stores increased from 273,700 jobs to 539,800 jobs, or 97.2 percent. Since 2007, the family clothing stores industry has lost 93,100 jobs, or 17.2 percent.

Fashion-related Occupations: Employment and Wages

In 2010, earnings in many occupations associated with apparel manufacturing were typically lower than the average for all occupations ($45,230). Among these occupations, fabric and apparel patternmakers—who use computer-aided design (CAD) software to determine the best layout of pattern pieces to minimize waste of material and to create a master pattern for each size within a range of garment sizes—earned an annual mean wage of $44,650. There were a total of 6,410 fabric and apparel patternmakers employed in 2010. Occupations such as textile and garment pressers, sewing machine operators, hand sewers, shoe and leather workers and repairers, and textile bleaching and dyeing machine operators and tenders earned a mean annual wage that was more than $15,000 below the average for all occupations. In 2010, sewing machine operators, with 142,860 workers, was the largest of these occupations.

Fashion designers earned an annual mean wage of $73,930 in 2010, over $25,000 more than the average for all occupations. There were a total of 16,010 fashion designers employed in 2010.

Fashion Designers

Fashion designers create original or exclusive custom-fitted clothing (e.g. haute couture), accessories, and footwear. In doing so, they must know how to sketch designs, select fabrics and patterns, and give instructions on how to make the products they design. Fashion designers work in wholesale or manufacturing establishments, apparel companies, retailers, theater or dance companies, and design firms.

Within the United States, most fashion designers work in large cities, such as New York or Los Angeles. In May 2010, almost 75 percent of all salaried fashion designers worked in New York and California. California led the nation, with a total of 4,480 employed fashion designers. Across the country, the mean annual wage earned by fashion designers ranged from $44,100 for those employed in Virginia to $80,650 for those employed in Maine and New York.

Among all states, California had the highest concentration of fashion designers. In general, location quotients are ratios that compare the concentration of a resource or activity, such as employment, in a defined area to that of a larger area or base. For example, location quotients can be used to compare State employment by occupation to that of the nation.

Fashion-related Occupations: Employment Outlook

Over the 2010–2020 period, as clothing continues to be made in other countries and the demand for custom clothing keeps declining, occupations such as sewing machine operators, fabric and apparel patternmakers, textile and garment pressers, and textile knitting and weaving machine setters, operators, and tenders are all projected to decrease in employment. Among those occupations, the number of sewing machine operators is expected to decline by 25.8%, or 42,100 jobs.

Employment in skilled occupations such as fashion designers and tailors, dressmakers, and custom sewers are projected to experience limited growth over the 2010–2020 period. Tailors, dressmakers, and custom sewers are projected to increase by 2.0 percent, or 900 workers, while fashion designers are projected to experience little or no change.

Mass Layoffs

From 1996 to 2011, the U.S. apparel manufacturing industry experienced many job losses—averaging 323 mass layoff events per year. During that period, the largest number of mass layoff events occurred in 1996, when the apparel manufacturing industry initiated a total of 706—leading to the filing of 67,511 initial claims for unemployment insurance benefits.

From 1996 to 2011, textile mills averaged a total of 200 mass layoff events per year, while leather and allied product manufacturers averaged 54 events per year. In 1996, apparel, textile mill, and leather and allied product manufacturers initiated a total of 1,040 mass layoff events—representing 7.1 percent of all mass layoff events in nonfarm establishments.

A mass layoff event occurs when fifty or more initial claims for unemployment insurance benefits are filed against an employer during a 5-week period, regardless of the duration of the layoff.

Injury and Illness Rates

A comparison of fashion-related industries shows that the rate of injuries varied among industries in 2010. Employees in thread mills had a higher than average injury rate of 6.7 per 100 full-time workers, whereas employees in yarn texturizing, throwing, and twisting thread mills suffered fewer injuries and illnesses at 1.8 percent.

In apparel manufacturing, the injury and illness rates in glove and mitten manufacturing, at 8.8 percent, was the highest of all measured occupations related to the fashion industry. Men's footwear (except athletic) manufacturing had a rate of 7.6 percent, compared with other footwear manufacturing at 3.6 percent.

Productivity

Productivity, a key measure of efficiency, is the amount of output produced per hour of work. Labor productivity in the U.S. manufacturing sector more than doubled from 1987 to 2010. Labor productivity also more than doubled over that period in U.S. textile mills and nearly doubled in footwear manufacturing. Labor productivity in apparel manufacturing followed a different pattern;

it grew at about the same rate as overall manufacturing productivity from 1987 to 2000 but generally declined from 2000 to 2010.

U.S. manufacturing output was nearly 50 percent higher in 2010 than in 1987 after adjusting for inflation, but real output in U.S. textile, apparel, and footwear manufacturing, declined substantially over the 1987–2010 period.

The number of hours that U.S. manufacturing employees worked remained fairly steady from 1987 to 2000 and then declined by about one-third between 2000 and 2010. Hours worked in U.S. textile, apparel, and footwear manufacturing declined nearly continuously and much more sharply than overall manufacturing hours during the 1987–2010 period.

Unit labor costs describe the relationship between compensation and labor productivity. Increases in hourly compensation increase unit labor costs; increases in labor productivity lower unit labor costs. Unit labor costs in U.S. manufacturing have held fairly steady since the late 1980s, meaning that manufacturers generally have been able to offset increases in compensation costs with improved efficiency. Unit labor costs for U.S. textile manufacturers also have held fairly steady since the late 1980s, but unit labor costs in U.S. apparel and footwear manufacturing were substantially higher in 2010 than in 1987.

Consumer Prices in the Apparel Industry

The Consumer Price Index for all items has risen at a much steeper rate than the indexes for apparel and shoes since 1978. Prices for apparel rose 62 percent from 1978 to 1998, declined somewhat through 2005, and have been fairly steady in recent years. Prices for footwear followed a similar pattern as apparel from 1978 to 2004, and footwear prices have increased somewhat more rapidly since 2004.

Consumer prices for men's and boys' apparel rose at somewhat faster rate than prices for women's and girls' apparel from 1978 to 1998. Prices for both categories declined somewhat through 2007 before leveling off in recent years. Prices for men's and women's footwear followed similar patterns as prices for apparel.

Consumer prices for infants' and toddlers' apparel rose about 69 percent from 1978 to 2000 and have generally declined since then.

Producer Prices in Apparel-related Industries

When shopping for clothing, shoes, and accessories in retail stores or over the Internet, a consumer's first thought about price is most likely not about the price exchange that occurs before the item is available at the retail level, although that transaction heavily influences the price the consumer sees. While producer prices for selected fashion-related industries have trended higher since December 2003, the Producer Price Index (PPI) for fabric mills, a major component in textile-related production, increased significantly from October 2010 until September 2011. In comparison, producer price increases for other industries such as footwear manufacturing and for accessories and other apparel were more muted until December 2011, when their rates of increase started to accelerate.

Import Prices in Apparel-related Industries

Have you ever wondered about the journey your clothes, shoes, and accessories traveled before these items found a home in your closet? Chances are your wardrobe includes many import components from across the globe. From December 2010 to February 2011, import prices for fabric mill products increased sharply and have continued to increase. Import prices for apparel accessories and other apparel manufacturing were higher than footwear manufacturing from September 2007 until October 2011, when footwear prices overtook apparel accessories and other apparel and have remained steady.

Compensation for U.S. and Foreign Apparel Manufacturers

In 2007, among those countries studied by the Bureau of Labor Statistics, Germany had the highest hourly compensation costs within the apparel manufacturing industry. The Philippines, with compensation costs at 88 cents per hour, had the lowest among those countries studied.

From 2006 to 2007, with the exception of Taiwan and Japan, hourly compensation costs increased in all countries studied—including the United States. From 2002 to 2007, Argentina and Australia experienced the largest increase in hourly compensation costs—increasing 154 percent. Over that period, Japan experienced the smallest increase in hourly compensation costs—from $11.77 per hour to $12.70 per hour, or 8 percent. Compensation costs for the United States increased from $15.37 per hour to $20.42 per hour, or 33 percent.

Source

Bureau of Labor Statistics (June 2012). "Spotlight On Statistics: Fashion. http://www.bls.gov/spotlight/2012/fashion/.

Statisticians can get enjoyment and inspiration from statistics for any of various reasons, as this book has shown, including the reason discussed in this essay—suddenly realizing that most of the core formulas in a basic statistics course share a common, unifying element that reduces their complexity and thus makes them easier to understand and use.

EPILOGUE

The Joy of Perceiving that Basic Statistics Is a "Mean World"

by

Richard Altschuler

As you have seen from reading this book, many statisticians, as well as scientists in diverse disciplines, philosophers, and others have experienced great joy from working with statistics, and some have used statistics to affect profound changes in society.

In this final essay, I would like to provide a personal anecdote about the joy I experienced one day while working with statistics, when I suddenly realized there is a *key* or "Rosetta Stone," one might say, for deciphering and simplifying most of the formulas taught in a basic statistics course and that are found in almost every introductory statistics textbook. And this key was "hiding right there in plain sight"—in the formulas themselves!

My "aha!"—which I playfully called a "*statori*" at the time (after the "zen" term *satori*, for an "epiphany" or breakthrough to "higher consciousness")—came after more than twenty years of privately tutoring hundreds of college students in statistics and teaching statistics courses to undergraduate and graduate students at Temple University and New York University. I can see now, in retrospect, that I have those students to thank for my revelation—especially the most "statistically challenged" of them—because *they* forced me to keep working, struggling to simplify the field, to explain the seemingly complex and varied formulas and concepts in the most easy-to-understand, everyday language possible.

What I suddenly realized one day is that the most frequently taught basic statistical formulas contain a *single statistical concept* or expression that *unifies* them, gives them a *commonality*, and thereby reduces their complexity and the difficulty of trying to learn, understand, and use them. That concept is the *arithmetic mean*, or as it is commonly called, the *"average"*!

Statistics textbooks, websites, and software packages equate the arithmetic mean with the "average" and define it the same way, in about the same words. The statistical software package SPSS (2013), for example, says the arithmetic mean "is commonly called the average"; The Encyclopedia Britannica Online (2013) states "The arithmetic mean (usually synonymous with average) represents a point about which the numbers balance"; and traditional textbooks simply state, respectively, "The arithmetic mean is an average" (Levin, 1978, p. 45) and the arithmetic mean "is doubtless the most familiar average; in fact, in common usage, it is often referred to as 'the average'" (Hamburg, 1979, p. 48).

As for the definition of the mean or average, it is universally defined as the sum of the scores of a set of observations divided by the total number of cases involved.

When this definition is expressed as a formula, the mean or average is simply sum/n. In this formula, as stated above, the "sum" (in the numerator) is the total value of two or more numbers (scores, values, etc.) added together, and "n" (in the denominator) is the number of "cases" (individuals, groups, etc.) that go into making up the sum. For example, if 5 different students have the ages 17, 18, 20, 23, and 25, then their total age is 103 years (the sum). So their mean or average age is expressed as 103/5. When you divide this through you get 20.6 years—their mean or average age.

Now, I suspect it may sound extraordinary to you—as it did to me, upon first realization—that a simple concept like the mean or average should solely or greatly define the most commonly taught and frequently used statistical formulas taught in a first-year statistics course—especially given the reputation statistics has of being difficult—but it is true! This single concept pervades the formulas of the variance, the mean deviation, the standard deviation, the standard error of the mean, the weighted mean, the correlation coefficient (or Pearson "r"), the average rate of change (the slope), the expected value, the t-test, the z-test, and analysis of variance (ANOVA), among about a dozen other core statistics that compose the basic statistics course.

This simple fact—about most of the core statistical formulas having a "mean nature"—is not only evidenced in their respective formulas, but also stated in their verbal definitions, which I have culled from prominent textbooks and websites and will now present. When reading them, do not be concerned if you do not recognize or understand all the terms or concepts involved, because the main point here is simply to show you that the important formulas in a basic statistics course are averages or mainly composed of one or more averages:

- the *variance* is defined as *the average* of the squared differences from the mean (note that in this definition, there are actually two means or averages—since the variance is an average that is obtained by subtracting values from "the mean," i.e., from the simple arithmetic mean or average, discussed above).

- the *standard deviation* is defined as the square root of the *variance* (note that in this definition, since the variance is an average, to get the standard deviation, all we do is "unsquare"—take the square root of—the variance).

- the *standard error of the mean* is defined as a particular type of standard deviation (i.e., of the sampling distribution of a statistic)—and we have already seen that the standard deviation is merely the square root of an average, the variance.

- the *correlation coefficient* (or *Pearson "r"*) is defined as *the average* of cross-products (also called a covariance) that are standardized by dividing through by both standard deviations.

- the *slope* is defined as *the average rate of change* of one variable in terms of another variable, e.g., income in terms of years of education.

- *analysis of variance* (*ANOVA*) is defined as a test of the equality of three or more *means* at the same time by using *variances* (specifically, by dividing the "between variance" by the "within variance").

- the *expected value* is defined as a *weighted average* of possible outcomes that can be thought of as the *arithmetic mean* of a set of numbers generated in exact proportion to their probability of occurring.

- the *grand mean* of a normal sampling distribution is defined as the *average* of the different *sample means* drawn from a population.

- the *t-test* for the difference between *two sample means* is defined as a test to determine if the sample means are significantly different or may be attributed to random chance, and is done by subtracting the *means* from one another and dividing the difference by the *standard errors* of the two samples (recall that the standard error is a particular type of standard deviation, and that the standard deviation is the square root of the variance, which is an average).

This list could go on to include other statistics that compose the core course, but you can see from the above just how much the mean pervades the basic statistical formulas. In this context, it should be mentioned that in addition to formulas being composed of means, in some formulas a single value or score (such as one person's age in a group of people) is subtracted from the mean of the group (e.g., to calculate the z-score); and that in other formulas the sample size ("n") is reduced by "1" or "2," depending on the number of samples involved in an analysis, among other slight variations in the core statistical formulas. But the basic idea remains the same: The mean or average solely or largely defines the most widely used formulas taught in an introductory statistics course.

Given this unifying feature of most of the basic statistical formulas, the task for statisticians is, of course, is to understand what makes each formula unique, so as to be able to use them appropriately; interpret them in prose when reporting the results of statistical analyses to others; and understand statistical findings reported in professional studies, government reports, and popular periodicals, on the Internet, or when hearing statistics reported over the TV and radio.

In other words, most of the statistics that compose the core course are basically "variations on a theme"—the theme being the average or arithmetic mean. If you focus on the *theme* (or commonalities) rather than on the *variations* (or differences), then I strongly believe, based on my experience as a statistics teacher, that you will see the field of statistics as less complex and, therefore, find it easier to learn and use statistics.

Looking at the statistical formulas this way, in terms of their "mean nature," requires a shift in perspective from the usual way statistics are presented in textbooks, which emphasizes the differences between and among the different formulas.

To illustrate how a *shift in perspective*—from difference to similarity, in this case—can simplify the understanding of a subject (even though the subject matter itself does not change, i.e., the statistical formulas remain the same), I would like you to consider the following example drawn from the study of biology: Imagine that a new biology student is studying cows, pigs, and sheep, and sees them all as *different* animals, focusing on their *unique* characteristics, such as their different sizes, colors, skeletal coverings (skin, fur, wool) and the different sounds they make, with the cow mooing, the sheep baahhing, and the pig oinking.

Now imagine that this student all of a sudden focuses on the animals' *similarities* instead of their differences, and—eureka!—he or she sees that cows, pigs, and sheep are all *mammals*! They all give birth to their young live, for example, and they all have the same organs in about the same places, walk on four legs, and function biologically about the same way, among so many other basic similarities. All of a sudden, these animals—cows, pigs, and sheep—seem so *alike;* and while their differences persist, they can all be thought of, conceived, and understood as "variations on a theme"—the theme of "mammal," in this case.

By focusing on what is the *same* about the different looking mammals, our hypothetical student doesn't *eliminate* their differences, of course, but the *change in viewpoint—from difference to similarity—simplifies* the understanding of the different animals and makes it much easier to learn about "how they tick" and behave. In still other words, *they* (the animals) don't change when *the viewpoint* of them changes—but the shift in focus makes a big difference in how easy or difficult it is to see or comprehend the wonderful, variegated world of animals we classify as mammals.

The same is true for understanding many different types of phenomena—such as plant life, computers, and foods. Once you see the "underlying principle" that governs their seemingly endless variety, you can understand a phenomenon much easier and gain mastery of it quickly, rather than be confused, frightened, and anxious by the seemingly endless variation.

If we further extend our example of "mammals" to basic statistics, we could say that most of the basic formulas are a type of "mammal"—only in our case we need to replace the concept of "mammal," of course, with the concept of "average" or "mean." Just as "mammal" is what unifies the apparently different animals we call cows, sheep, and pigs, so "mean" or "average" is what unifies most of the statistical formulas at the heart of the basic statistics course.

In concluding this essay, I would like to convey my belief that, just as I got pleasure from my "aha" about the "mean nature" of the basic statistics, and also from sharing that viewpoint with the students I taught, you also have the ability see something unique or special about the statistical formulas, concepts, or applications that brings you pleasure and, perhaps, inspires you to want to use statistics for both your own benefit and in ways that have a positive impact on individuals and the world.

Source

This essay was originally written for the anthology by the editor.

References

Encyclopedia Britannica Online. (2013). http://www.britannica.com/EBchecked/topic/371524/mean

Hamburg, Morris. (1979). *Basic statistics*. New York: Harcourt Brace Jovanovich, Inc.

Levin, Richard. (1978). *Statistics for management*. Englewood Cliffs, N.J.: Prentice-Hall.

SPSS. (2013). http://www.ats.ucla.edu/stat/spss/output/descriptives.htm

A Potpourri of Quotations about Statistics and Statisticians

Humor/Satire/Whimsy

"Satan delights equally in statistics and in quoting scripture."
— H. G. Wells

"The statistics on sanity are that one out of every four Americans is suffering from some form of mental illness. Think of your three best friends. If they're okay, then it's you."
— Rita Mae Brown

"Statistics show that of those who contract the habit of eating, very few survive."
— George Bernard Shaw

"The first time I was in a statistics course, I was there to teach it."
— John Tukey

The 50-50-90 rule: Anytime you have a 50-50 chance of getting something right, there's a 90% probability you'll get it wrong.
— Andy Rooney

"Statistics are somewhat like old medical journals, or like revolvers in newly opened mining districts. Most men rarely use them, and find it troublesome to preserve them so as to have them easy of access; but when they do want them, they want them badly."
—John Shaw Billings

"Like dreams, statistics are a form of wish fulfillment."
— Jean Baudrillard

"In God we trust. All others must bring data."
— W. Edwards Deming

Probability/Uncertainty/Chance

"The probability that we may fail in the struggle ought not to deter us from the support of a cause we believe to be just."
— Abraham Lincoln

"The most important questions of life are, for the most part, really only problems of probability."
— Simon Laplace

"The essence of life is statistical improbability on a colossal scale."
—Richard Dawkins

"I think it is much more interesting to live with uncertainty than to live with answers that might be wrong."
— Richard Feynman

"The true logic of this world is in the calculus of probabilities."
— James Clerk Maxwell

"A statistical analysis, properly conducted, is a delicate dissection of uncertainties, a surgery of suppositions."
— M. J. Moroney

"My thesis is simply this: probability does not exist."
— Bruno de Finetti

"Statistically, the probability of any one of us being here is so small that you'd think the mere fact of existing would keep us all in a contented dazzlement of surprise."
— Lewis Thomas

"Life is a school of probability."
— Walter Bagehot

"The new always happens against the overwhelming odds of statistical laws and their probability, which for all practical, everyday purposes amounts to certainty; the new therefore always appears in the guise of a miracle."
— Hannah Arendt

"The excitement that a gambler feels when making a bet is equal to the amount he might win times the probability of winning it."
— Blaise Pascal

"We have big brains, but we live in an incomprehensibly large universe. The virtue in thinking probabilistically is that you will force yourself to stop and smell the data—slow down, and consider the imperfections in your thinking. Over time, you should find that this makes your decision making better."
— Nate Silver

"Probability is the very guide of life."
— Marcus Tullius Cicero

"It is very certain that, when it is not in our power to determine what is true, we ought to act according to what is most probable."
— René Descartes

"The million, million, million . . . to one chance happens once in a million, million, million . . . times no matter how surprised we may be that it results in us."
— Ronald Aylmer Fisher

"All knowledge resolves into probability."
— David Hume

"The investigation [of probabilities] is one that deserves the attention of philosophers in showing how in the final analysis there is a regularity underlying the very things that seem to us to pertain entirely to chance, and in unveiling the hidden but constant cause on which that regularity depends."
—Simon Laplace

"People are entirely too disbelieving of coincidence. They are far too ready to dismiss it and to build arcane structures of extremely rickety substance in order to avoid it. I, on the other hand, see coincidence everywhere as an inevitable consequence of the laws of probability, according to which having no unusual coincidence is far more unusual than any coincidence could possibly be."
— Isaac Asimov

"You believe in a God who plays dice, and I in complete law and order in a world which objectively exists, and which I, in a wildly speculative way, am trying to capture. I firmly believe, but hope that someone will discover a more realistic way, or rather a more tangible basis than it has been my lot to do. Even the great initial success of the quantum theory does not make me believe in the fundamental dice game, although I am well aware that your younger colleagues interpret this as a consequence of senility."
—Albert Einstein

"Reality dishes out experiences using probability, not plausibility."
—Eliezer S. Yudkowsky

"I could prove God statistically. Take the human body alone—the chances that all the functions of an individual would just happen is a statistical monstrosity."
—George Gallup

"While the individual man is an insoluble puzzle, in the aggregate he becomes a mathematical certainty. You can, for example, never foretell what any one man will be up to, but you can say with precision what an average number will be up to. Individuals vary, but percentages remain constant. So says the statistician."
—Arthur Conan Doyle

"In the Bayesian worldview, prediction is the yardstick by which we measure progress. We can perhaps never know the truth with 100 percent certainty, but making correct predictions is the way to tell if we're getting closer."
— Nate Silver

Statisticians

"Statisticians, like artists, have the bad habit of falling in love with their models."
— George Box

"The best thing about being a statistician is that you get to play in everyone's backyard."
— John Tukey

"When I see articles with lots of significance tests, I say that the statisticians are p-ing on the research."
— Herman Friedmann

"I keep saying that the sexy job in the next 10 years will be statisticians. And I'm not kidding."
— Hal Varian

"... the statistician knows ... that in nature there never was a normal distribution, there never was a straight line, yet with normal and linear assumptions, known to be false, he can often derive results which match, to a useful approximation, those found in the real world."
— George Box

"We statisticians, as a police of science (a label some dislike but I am proud of . . .), have the fundamental duty of helping others to engage in statistical thinking as a necessary step of scientific inquiry and evidence-based policy formulation. In order to truly fulfill this task, we must constantly firm up and deepen our own foundation, and resist the temptation of competing for "methods and results" without pondering deeply whether we are helping others or actually harming them by effectively encouraging more false discoveries or misguided policies. Otherwise, we indeed can lose our identity, no matter how much we are desired or feared now."
— Xiao-Li Meng

"I went to parties and heard a little groan when people heard what I did. Now they're all excited to meet me. It's not because of a new after-shave. Arcane statistical analysis, the business of making sense of our growing data mountains, has become high tech's hottest calling."
— Robert Tibshirani

General

"To understand God's Thoughts we must study statistics for these are the measure of His purpose."
— Florence Nightingale

"Whether statistics be an art or a science . . . or a scientific art we concern ourselves little. It is the basis of social and political dynamics, and affords the only secure ground on which the truth or falsehood of the theories and hypotheses of that complicated science can be brought to the test."
—Adolphe Quetelet

"Statistics are "the only tools by which an opening can be cut through the formidable thicket of difficulties that bars the path of those who pursue the science of man."
— Sir Francis Galton

". . . the problem of evolution is a problem in statistics, in the vital statistics of populations."
—Karl Pearson

"This is the age of statistics, Mr. Speaker. . . . The legislator without statistics is like the mariner at sea without the compass. Nothing can safely be committed to his guidance."
—James A. Garfield

"I've come loaded with statistics, for I've noticed that a man can't prove anything without statistics. No man can"
—Mark Twain

"Statistical methods are essential to social studies, and it is principally by the aid of such methods that these studies may be raised to the rank of sciences."
— Ronald A. Fisher

"Many of the *Moneyball*-era debates concerned not *whether* statistics should be used, but *which* ones should be taken into account."
— Nate Silver

"We all know that Americans love their statistics—in sport, obviously. And in finance too."
— Evan Davis

"The science of statistics is the chief instrumentality through which the progress of civilization is now measured, and by which its development hereafter will be largely controlled."
— S. N. D. North

"Statistics are the heart of democracy."
—Simeon Strunsky

"Conducting data analysis is like drinking a fine wine. It is important to swirl and sniff the wine, to unpack the complex bouquet and to appreciate the experience. Gulping the wine doesn't work."
— Daniel B. Wright

"It is the mark of a truly intelligent person to be moved by statistics."
— George Bernard Shaw

"As much as it pleases me to see statistical data introduced in the Supreme Court, the act of citing statistical factoids is not the same thing as drawing sound inferences from them."
—Nate Silver

Full and minute statistical details are to the lawgiver, as the chart, the compass, and the lead to the navigator.
—Lord Brougham

"It can be stated without exaggeration that more psychology can be learned from statistical averages than from all philosophies, except Aristotle."
—William Wundt

"... the social sciences in general and social statistics in particular have a great service to render to government and through government to mankind."
—Wesley C. Mitchell

"What is statistics? To many, statistics is the class they took in college or figures on the sports pages. But statistics is so much more. Statistics is the science of learning from data and measuring, controlling, and communicating uncertainty."
—Marie Davidian

"Statistical thinking will one day be as necessary for efficient citizenship as the ability to read and write."
— H. G. Wells

Works Cited

The following is a comprehensive list of the works cited in the readings, presented in alphabetical order by last name of author.

A report of the Surgeon General: Preventing tobacco use among youth and young adults: a report of the Surgeon General, 2012. (2012). Atlanta (GA): US Department of Health and Human Services, Centers for Disease Control and Prevention, National Center for Chronic Disease Prevention and Health Promotion, Office on Smoking and Health.

Bayes, Thomas. (1763). *An Essay towards solving a problem in the doctrine of chances.* Philosophical Transactions:1683-1775. The Royal Society. http://www.jstor.org/stable/105741.

Beirne, Piers. (1993). *Inventing criminality: Essays on the rise of 'Homo Criminalis.'* Albany, NY: State University of New York Press.

Belin TR, Fischer HJ, Zigler CM. (2011). Using a density-variation/compactness measure to evaluate redistricting plans for partisan bias and electoral responsiveness. *Statistics, Politics, and Policy*, Volume 2: Article 3.

Bernhardt, Victoria L. (2000). New routes open when one type of data crosses another. *Journal of Staff Development 21*, 1 (Winter). http://www.nsdc.org/library/jsd/bernhardt211.html

Bernhardt, Victoria L. (2001). Intersections. *Journal of Staff Development, 21*(1), 33-36 (Winter). (EJ 600 392) Available at: http://www.nsdc.org/library/jsd/bernhardt211.html

Best E., Josie, G., & Walker, C. (1961). A Canadian study of mortality in relation to smoking habits, a preliminary report. *Canad J Pub Health, 52*:99-106.

Biometrika. (1901). Vol. 1, No. 1, Oct. Biometrika Trust. http://www.jstor.org/discover/10.2307/2331669?uid=3739560&uid=2134&uid=2&uid=70&uid=4&uid=3739256&sid=21102670168653

Bureau of Labor Statistics (2012, June). Spotlight On Statistics: Fashion. http://www.bls.gov/spotlight/2012/fashion/

Calhoun, Emily F. (1994). *How to use action research in the self-renewing school.* Alexandria, Virginia: Association for Supervision and Curriculum Development.

Centers for Disease Control and Prevention (CDC) Smoking in top-grossing movies, United States 2010. (2011). *MMWR Morb Mortal Wkly Rep 60*(27): 910–3.

Chrispeels, Janet H., and others. School leadership teams: A process model of team development. (2000). *School Effectiveness and School Improvement 11*, 1, March: 20-56. EJ 611 222.

Clark, Kenneth B. & Clark, Mamie, P. (1947). Racial identification and preference in Negro children. Condensed by the authors from an unpublished study. http://i2.cdn.turner.com/cnn-/2010/images/05/13/doll.study.1947.pdf

Congressional Record. (2013). May 21, page S3662. http://-thomas.loc.gov/cgi-bin/query/z?c113:S.RES.150

Cook, Sir Edward. (1913). *The life of Florence Nightingale, in two volumes, vol. I (1820–1861).* London: Macmillan and Co., Limited.

Davidian, Marie. (2013). 2013: The international year of statistics. *Huff Post: Science.* http://www.huffingtonpost.com-/marie-davidian/2013-the-international-ye_b2670704.html

Davidson, Max. (2010). Bill Bryson: Have faith, science can solve our problems. *Daily Telegraph* (26 September).

Debow, D. B. (1854). *Statistical view of the United States, embracing its territory, population—white, free colored, and slave-moral and social condition, industry, property, and revenue; the detailed statistics of cities, towns, and counties; being a compendium of the seventh census, to which are added the results of every previous census, beginning with 1790, in comparative tables, with explanatory and illustrative notes, based upon the schedules and other official sources of information.* Washington: Beverly Tucker, Senate printer.

Doll, R. & Hill, A. (1956). Lung cancer and other causes of death in relation to smoking. *Brit Med J.* 2:1071-1081.

Du, Yi, & Larry Fuglesten. (2001). Beyond test scores: Edina public schools' use of surveys to collect school profile and accountability data. *ERS Spectrum* (Summer): 20-25.

Dunn, J. Jr., Linden, G. & Breslow, L. (1960). Lung cancer mortality experience of men in certain occupations in California. *Am J Pub Health. 50*:1475-1487.

Encyclopedia Britannica Online. (2013). Entry for the mean. http://www.britannica.com/EBchecked/topic/371524/mean.

Ending the tobacco epidemic: A tobacco control strategic action plan for the US Department of Health and Human Services. Washington (DC): US Department of Health and Human Services; 2010.

Feldman, Jay, & Rosann Tung. (2001). Using data-based inquiry and decision making to improve instruction. *ERS Spectrum* (Summer): 10-19.

Galton, Francis. (1894). *Natural inheritance.* New York and London: Macmillan and Co.

Galton, Francis. (1883). *Inquiries into human faculty and its development.* New York and London: Macmillan and Co.

Garfield, James A. (1869). *The Congressional Globe.* UNT Digital Library. http://digital.library.unt.edu/ark:/67531/metadc30-883/m1/550/sizes/?q=achenwall

Getstats. (2012). Higgs Boson and the statistics of certainty. http://www.getstats.org.uk/2012/07/04/higgs-boson-and-the-statistics-of-certainty/

Geliebter A. (1988). Gastric distension and gastric capacity in relation to food intake. *Physiology and Behavior, 44*:665-668.

Geliebter A, Schachter S, Lohmann C, Feldman H, & Hashim SA. (1996). Reduced stomach capacity in obese subjects after dieting. *Am J Clin Nutr 63*:170-173.

Generation Next: Interview with Francesca Dominici. *Johns Hopkins Public Health* (2003). http://magazine.jhsph.edu/2003/-fall/generation_nxt/dominici.html

Glantz SA, Iaccopucci A, Titus K, & Polansky JR. (2012). Smoking in top-grossing US movies, 2011. *Prev Chronic Dis 9*:E150. doi: 10.5888/pcd9.120170.

Goldberg, Joseph P. & Moye, William T. (1985). *The first hundred years of the Bureau of Labor Statistics* (1884-1984). Washington, DC: U. S. Government Printing Office. http://www.-bls.gov/opub/blsfirsthundredyears/`00_years_of_bls.pdf).

Hamburg, Morris. (1979). *Basic statistics.* New York: Harcourt Brace Jovanovich, Inc.

Hammond, E. & Horn, D. (1958). Smoking and death rates—reported on forty-four months of follow-up on 187,783 men. *JAMA*; *166*:1159-1172, 1294-1308.

Hankins, Frank Hamilton. (1908). *Adolphe Quetelet as statistician.* New York: Columbia University, doctoral dissertation.

Hardy, Quentin (2012). What are the odds that stats would be this popular? *New York Times.* http://bits.blogs.nytimes.com/2012/01/26/what-are-the-odds-that-stats-would-get-this-popular/

Holcomb, Edie L. (1999). Getting *excited about data: How to combine people, passion, and proof.* Thousand Oaks, California: Corwin Press.

Jackman, Simon. (2012). Pollster predictions: 91.4% chance Obama wins, 303 or 302 EVs. http://www.huffingtonpost.com/simon-jackman/pollster-predictions_b_2081013.html

Killion, Joellen, & G. Thomas Bellamy. (2000). On the job: data analysts focus school improvement efforts. *Journal of Staff Development 21,* 1 (Winter). http://www.nsdc.org/library/jsd/killion211.html

Lachat, M.A. (2002). *Data-driven high school reform: The breaking ranks model.* Hampton, NH: Center for Resource Management.

Laplace, Pierre-Simon. (1825; 1995). *Philosophical essay on probabilities.* Trans. By Andrew I. Dale. NY: Springer-Verlag.

Levin, Richard. (1978). *Statistics for management.* Englewood Cliffs, N.J.: Prentice-Hall.

Lewis, Michael. (2003). *Moneyball: The art of winning an unfair game.* New York: W. W. Norton & Company.

Lohr, Steve. (2009). For today's graduate, just one word: Statistics. *New York Times* http://www.nytimes.com/2009/08/06/-technology/06stats.html?_r=1&

Lohr, Steve. (2012). The age of big data. *New York Times*: http://www.nytimes.com/2012/02/12/sunday-review/big-datas-impact-in-the-world.html?pagewanted=all&_r=1&

Manyika, J. et al. (2011). Big data: The next frontier for innovation, competition, and productivity. *Insights and Publications: McKinsey Global Institute.* http://www.mckinsey.com/insights/business_technology/big_data_the_next_frontier_for_innovation

Marx, Karl. (1867; 1978). *Capital*, volume one, edited by Robert C. Tucker. New York: Norton.

McAfee, T., MD, MPH & Tynan, M. (2012). Smoking in movies: A new Centers for Disease Control and Prevention core surveillance indicator. *Prev Chronic Dis*; 9:120261.DOI: http://-dx.doi.org/10.5888pcd9.120261.

McLean, James E. (1995). Improving *education through action research: A guide for administrators and teachers.* Thousand Oaks, California: Corwin Press

Meitzen, August (1891). *History, theory and technique of statistics.* Trans. by Roland Falkner. Philadelphia: American Academy of Political and Social Science.

Miller, A. Christine. (2000). School reform in action. Paper presented to the American Educational Research Association Conference, New Orleans, April 28.

Millett C, & Glantz SA. (2010). Assigning an "18" rating to movies with tobacco imagery is essential to reduce youth smoking. *Thorax*; 65(5):377–8. doi: 10.1136/thx.2009.133108.

Mitchell, Wesley C. (1919). Statistics and government. *American Statistical Association, New Series, No. 125*, 223-235.

Moneyball: The art of winning an unfair game, summary on Wikipedia. (2013). http://en.wikipedia.org/wiki/Moneyball

National Cancer Institute tobacco control monograph 19: The role of the media in promoting and reducing tobacco use. (2008). Bethesda (MD): US Department of Health and Human Services, National Institutes of Health, National Cancer Institute.

National Institutes of Health. Advisory Committee to the Director. (2011). http://acd.od.nih.gov/working-groups.htm

North Central Regional Educational Laboratory. (2000). *Using data to bring about positive results in school improvement efforts.* Oakbrook, Illinois.

Pearson, Karl. (1924). *The life, letters and labours of Francis Galton*, vol. 2. London: Cambridge University Press.

Perlmutter, S. et al. (1999). Measurements of Ω and Λ from 42 high-redshift supernovae. http://iopscience.iop.org/0004-637X/-517/2/565/

Porter, Theodore. (1986). *The rise of statistical thinking: 1820—1900*. Princeton, NJ: Princeton University Press.

Price, Richard. (1763). Introduction to *An essay towards solving a problem in the doctrine of chances,* by Thomas Bayes, Nov. 10. Philosophical Transactions: 1683-1775. The Royal Society. http://www.jstor.org/stable/105741.

Quetelet, Adolphe. (1835). *Essai de physique sociale*. Paris: Bachelier.

Quetelet, Adolphe. (1842). *A treatise on man and the development of his faculties*. Reprinted with an introduction by Solomon Diamond (1969). Gainesville: Scholars' Facsimiles and Reprints.

Rehmeyer, Julie. (2009). Darwin: The reluctant mathematician. *Science News, February* 11. http://www.sciencenews.org/view/generic/id/40740/description/Darwin_The_reluctant_mathematician_

Ridpath, John Clark. (1881). *The life and work of James A. Garfield*. Cincinnati, OH: Jones Brothers. http://archive.org/stream/lifeandworkjame02ridpgoog/lifeandworkjame02ridpgoog_djvu.txt

Sargent JD, Tanski SE, & Gibson J. (2007). Exposure to movie smoking among US adolescents aged 10 to 14 years: A population estimate. *Pediatrics*;*119*(5):e1167–76. doi: 10.1542/peds.2006-2897

Sargent JD, Tanski S, & Stoolmiller M. (2012). Influence of motion picture rating on adolescent response to movie smoking. *Pediatrics 130*(2):228–36. doi: 10.1542/peds.2011-1787.

Schwartz, Wendy. (2002). Data-driven equity in urban schools. Adapted from *ERIC Digest*, Identifier: ED467688. ERIC Clearinghouse on Urban Education New York NY. http://files.eric.ed.gov/fulltext/ED467688.pdf

Severo, Richard. (2005). Obituary of Kenneth Clark. *New York Times*, Tuesday, May 3. http://www.nytimes.com/2005/05/02/nyregion/02clark.html?_r=0

Silver, Nate. (2012). *The signal and the noise: Why so many predictions fail—but some don't*. New York: The Penguin Press.

Silver, Nate. (2013). The White House is not a metronome. *New York Times*. (July 18). http://fivethirtyeight.blogs.nytimes.com/

Slowinksi, J. (2002). Data-driven equity: Eliminating the achievement gap and improving learning for all students. Unpublished manuscript, Vinalhaven Schools, Vinalhaven, ME.

Smoking and health: A report of the advisory committee to the Surgeon General of the Public Health Services. (1964). Released by U.S. Surgeon General Luther Terry. http://profiles.nlm.nih.gov/ps/access/NNBBMT.pdf

SPSS. (2013). Definition of the arithmetic mean. http://www.ats.ucla.edu/stat/spss/output/descriptives.htm

Sparks, Dennis. (2000). Results are the reason. *Journal of Staff Development 21*,1 (Winter).

Teitell, Beth. (2012). The allure of the statistics field grows. *The Boston Globe*. http://www.bostonglobe.com/lifestyle/style/2012/11/21/

The International Year of Statistics (2013). http://www.statistics2013.org/

Tobacco use in healthy people 2020. (2010). Washington (DC): US Department of Health and Human Services.

United States Department of Agriculture/National Agricultural Statistics Service. (2009). Education and Outreach: Importance of Ag estimates. http://www.nass.usda.gov/Education_and_Outreach/Understanding_Statistics/Importance_of_Ag_Estimates/index.asp

United States Department of Labor, Bureau of Labor Statistics, Spotlight on Statistics. (2009). http://www.bls.gov/spotlight/2009/125_anniversary/home.htm

United States Supreme Court Footnote 11 Citations: Clark, K. B. Effect of prejudice and discrimination on personality development (Midcentury White House Conference on Children and Youth, 1950); Witmer and Kotinsky, *Personality in the making* (1952), c. VI; Deutscher and Chein, The psychological effects of enforced segregation: A survey of social science opinion, *26 J. Psychol.* 259-287 (1948); Chein, What are the psychological effects of segregation under conditions of equal facilities?, 3 *Int. J. Opinion and Attitude Res. 229* (1949); Brameld, *Educational costs, in discrimination and national welfare* (MacIver, ed., (1949), 44-48; Frazier, *The Negro in the United States* (1949), 674-681. and see generally Myrdal, *An American dilemma* (1944).

Votamatic: (2012). Forecasts and polling analysis for the 2012 presidential election. http://votamatic.org/forecast-detail/

Wade, Howard H., (2001). Data inquiry and analysis for educational reform. ED461911. ERIC Clearinghouse on Educational Management, http://files.eric.ed.gov/fulltext/ED461911.pdf

Wang GJ, Tomasi D, Backus W, Wang R, Telang F, Geliebter A, Korner J, Bauman A, Fowler JS, Thanos PK, & Volkow ND. (2008). Gastric distention activates satiety circuitry in the human brain. *Neuroimage*, Feb 15, 39(4):1824-31.

Wells/Wilks on statistical thinking. http://www.causeweb.-org/cwis/SPTFullRecord.php?ResourceId=1240

World Health Organization. (2009). *Smoke-free movies: From evidence to action.* Geneva (CH): World Health Organization. http://www.who.int/tobacco/smoke_free_movies/enAccessed 10/18/2012

DATE DUE